The Plant Contract

Critical Plant Studies

PHILOSOPHY, LITERATURE, CULTURE

Series Editor

Michael Marder
(*IKERBASQUE/The University of the Basque Country, Vitoria*)

VOLUME 3

The titles published in this series are listed at *brill.com/cpst*

The Plant Contract

Art's Return to Vegetal Life

By

Prudence Gibson

BRILL
RODOPI

LEIDEN | BOSTON

Cover illustration: Hayden Fowler. *Dark Ecology 2015*. Mixed media, CCTV cameras, 4-channel video, monitor. Courtesy of the artist.

Library of Congress Cataloging-in-Publication Data

Names: Gibson, Prudence, author.
Title: The plant contract : art's return to vegetal life / by Prudence Gibson.
Description: Leiden ; Boston : Brill, 2018. | Series: Critical plant studies:
  philosophy, literature, culture, ISSN 2213-0659 ; VOLUME 3 | Includes
  bibliographical references and index. |
Identifiers: LCCN 2017051187 (print) | LCCN 2017054739 (ebook) | ISBN
  9789004360549 (E-book) | ISBN 9789004353039 (pbk. : alk. paper)
Subjects: LCSH: Nature (Aesthetics) | Plants. | Plants (Philosophy) |
  Philosophy of nature. | Human ecology. | Art--Philosophy.
Classification: LCC BH301.N4 (ebook) | LCC BH301.N4 G53 2018 (print) | DDC
  113/.8--dc23
LC record available at https://lccn.loc.gov/2017051187

Typeface for the Latin, Greek, and Cyrillic scripts: "Brill". See and download: brill.com/brill-typeface.

ISSN 2213-0659
ISBN 978-90-04-35303-9 (paperback)
ISBN 978-90-04-36054-9 (e-book)

Contents

Acknowledgements

For this monograph, Douglas Kahn was a source of inspiration, intellectual friendship and structural advice. Thanks also to Douglas' daughter Aleisha Kahn for her excellent proofing. I want to thank Edward Scheer for his unconditional and undying support; and the friendship of colleagues Sigi Jottkandt, Paul Thomas, Andrew Murphie and Helen Groth. Thank you to all the artists who shared their work and ideas with me for this text. Finally, I have to thank my dear friend and collaborative writer Monica Gagliano whose work in the area of Critical Plant Studies has been a source of wonder and philosophical provocation for this art-focused version of plant theory.

Introduction

This book focusses on the art aesthetics of new plant discoveries. It proposes that contemporary visual art and culture can mitigate the way we have forgotten the natural world. Art can address the problems of no longer recognizing nature, by re-introducing us to it. This is more than a call for a return to nature and specifically to plants. It is also an effort to engage with the massively distributed and globalized issue of climate change, in a small and localized way, human by human, artist by artist, plant by plant. Humans will always see the bear not the woods, the lion not the grasslands, the camel not the desert dunes. We are attuned to see animals, and to see ourselves. This book aims to train our eyes and thoughts back to the landscape; it marks a return to the vegetal backdrop that becomes a multitude of subjectivities.

If we can 'see' the vegetal world once more, we might remember what we are. Plants are the key to curing vegetal blindness. What are these plants and how did we forget them? In partnership with art, within the larger field of aesthetics, plants are calling us back to the world. Somewhere, somehow, we have lost respect for the world, for nature and for each other. When I was a child, I disliked accompanying my father as he walked through the neighbourhood to collect kindling from the local park to make our evening winter fire. Why? Because he would stop to chat courteously with almost every person we came across. He had an uncanny knack of showing the greatest of respect to all who passed him by. He would also pause and bring my attention to a flowering camellia sasanqua or a spectacular squiggly gum tree trunk. Now I look back and see the virtue of taking time and showing courtesy.

Michel Serres, in his text *The Natural Contract*, mourns the lost pact among sailors on the high seas, where common courtesy and a code of conduct flourished. Serres promises, in his natural contract exposition, that 'the natural world will never again be our property, either private or common, but our symbiont.'[1] The earth, Serres explains, speaks to us in terms of bonds and pacts, forces and interactions, which require a contract.[2] It is an agreement amongst humans to honour the Earth, the world, to honour nature as our mutual symbiont.

This book is a deep endorsement of Serres' natural contract. The Plant Contract emerges directly from my conclusion that a contractual nature study requires further attention and that there is a place in Critical Plant Studies

---

1 Michel Serres, *The Natural Contract*, Ann Arbor: University of Michigan Press, 1995, 44.
2 Ibid, 39.

and alongside vegetal philosophy for such inquiry. It is also a gesture of respect for the genius of vegetal life itself,[3] via an investigation into nature rights, the wasteland, feminist plants, robotic technologies and the greatest artists working in this sphere of vegetal philosophy. It is an inquiry into how art can mediate and express plant-philosophy to a wider public. This is a phytological project based on plant-love. But it also acknowledges that some plants are wicked, some are defiant and others cannot be controlled. Respect and maybe even a little fear for plants reminds humans of our true place in the world.

Later in this introduction, after establishing the aesthetic (visual art) angle, this text elaborates the kinds of contractual emphases that the book chapters elaborate. However here, at the beginning of the beginning, it needs to be explained that climate change's inestimable scale, its vastness, is altogether too imposing for humans to consider or contend with. Humans have tread a course of dominion or sovereignty over nature which has contributed to the ecological damage we know today. We no longer understand the seasons, the weather or vegetal life around us. Only when our car gets bogged on a friend's farm or when heavy rain causes leaks inside the house or we start sneezing from spring pollen do we even notice that we live in a natural world. These days, we live indoors, indifferent to weather. We have neglected the world, we have turned our back on the crops, the farm animals and the harvests. We don't remember the plough or the horse's harness or the hay bales. We stare at our screens and keep typing, whether or not the weather is inclement.

A plant contract is a further development of the natural contract. It supports Serres' notion of a need to rediscover, however naively, the natural world. Serres notes that the social contract (presented in 1762 by Jean Jacques Rousseau) marked an abandonment of nature in order to create a social pact. Humans formed a governance structure that managed society fairly and equitably. From that moment, Serres suggests, we forgot our aforementioned nature, which is now distant and mute. A plant contract is an effort to un-mute nature, via plant science and art showing evidence that plants communicate to one another via emissions. If we stop talking, and stop typing, for brief periods, we may be able to hear the plants and to remember nature which has become so distant. If, via

---

3   Luce Irigaray refers to plant 'genius,' whereas Matthew Hall and Anthony Trewavas to plant 'intelligence': Matthew Hall 'Bridging the Gulf: Moving, Sensing, Intelligent Plants,' in *Plants as Persons: A Philosophical Botany*, Albany: SUNY Press, 2011; Anthony Trewavas, 'Aspects of Plant Intelligence,' *Annals of Botany*, Vol 92, No 1 2003, 1–20.

the social contract, we have cast off from nature, then the plant contract is a means of re-rooting (or re-mooring) us into the land.

The Plant Contract is an act of resuscitation, a breath of renewed life into a natural contract. We must avoid the behavior of the destroyer and the devourer, for this is parasitism. With our endless human desire to own property and to exert influence over the land, we have become intractable and reduced to human nature, rather than the human. The 'rights of mastery and property come down to parasitism. Rights of symbiosis however are reciprocal and must give back to nature now a legal subject.'[4] The plant contract is symbiosis, rejecting parasitism.

If a natural contract brought our attention to the collectivity and the totality of the world, the plant contract continues that tradition, by bringing attention to the totality of the vegetal world and the way in which art punctures that world to disrupt devastating human habits. This totality affords and encourages the power and forces of art. The impact of the aesthetic experience can assist a renewed view of the totality of the natural world.

The Plants in Art

My inquiry here is immersed in the realm of visual art. There is a cohort of performance, video, bio, sonic, environmental, installation and social engagement artists who are interpreting and experimenting with plant information in their artworks. This creates a mediation and a communication, as well as an expression, of vegetal thought. Yes, art raises awareness for critical issues but it also infiltrates our cultural and social being. Plant art has a greater capacity than mere aesthetic instrumentalism. Just as art has its own independent entelechy or energizing force, so too plants exist outside any human manipulation of them. This connection between plant and a flat art ontology (the collapse of any hierarchical structure) drives the ideas in this book and helps to begin a process of devising what we might call this plant contract, a new deal for the vegetal world: a means of altering our perception of nature by attempting to see all the parts, as well as the overall sum, of plant art.[5] As we discover more and more about the radical and uncanny life of plants, it is easy to be inspired

---

4   Michel Serres, *The Natural Contract*, Ann Arbor: University of Michigan Press, 1995, 38.

5   Dalia Nassar, 'Metaphoric Plants: Goethe's *Metamorphosis of Plants* and the Metaphors of Reason,' *Covert Plants*, Santa Barbara: Punctum Books, 2017.

by what this could mean for our existing contracts in social and political life. Greater care? A sustainable approach? Mutuality?

Nature can no longer be seen as an inert backdrop to human action. A tree does not call itself a tree: it is a complex organism of charging hormones, firing synapses, busy photosynthesis and strange excrescences. At the same time that plants have been modified, synthetically acclimatized, commercialised and commodified, word is spreading about how interesting and capable they are, for-themselves and in-themselves. It is not difficult to speculate that there may be a future collision between those that acknowledge the appearance of plants as ontologically independent entities, and those who rely on plants' agricultural and corporatized profit potential. We are moving into the epoch of the plant, where careless productions and practices relating to vegetal life are becoming more controversial. How this epoch will unfold relies upon how humans deal with plant world relations.

In the context of art, my interpretation of Critical Plant Studies is as a perception of various ecologies of the world, as interactions of energetic activity. Each ecology connects with the next and to its entirety as well. This is an aggregated system of addressing changes to notions of nature, away from transcendence, away from sublime aesthetics. Instead of the human mind thinking like a machine, in this text, the human mind approaches its environment like a plant system. At the very least, this book is the metaphoric gesture of attempting to do so. Vegetal studies can assist in understanding human connections with its ecologies, and human cultural practice can help humans to assist plant functions, in order to provide a more cooperative way of living. Bringing art practice into the mix provides an accelerating mediation. This is the purpose of combining art and plants within a critical context: the dovetail effect.

In an interview with Michael Marder on the relations between plants and art, he said,

> Art, in turn, is sublimated plant-sensing. Aesthesis, at the root of sensation and aesthetics, is not the exclusive province of animals and humans; as we know, plants are highly receptive to a variety of environmental factors, from light and moisture gradients to vibrations. To be sure, plants neither think nor see in images, but this does not mean that they neither think nor see.[6]

---

6   Prudence Gibson (ed). 'Interview with Michael Marder' in *Covert Plants*, Santa Barbara: Punctum Books, 2017.

Sublimated plant-sensing reminds us of the capacity of art to create reactive sensory experiences outside the everyday. The thinking, sensing and seeing of plants and of aesthetics are deeply connected, despite their apparent differences. Plants root, shoot, flower, photosynthesise and self-generate. Aesthetics creates, represents, expresses. The latter mobilises analogy and allegory and is a human-centred study of the nature of beauty and taste. The connection lies in the deep beauty of considering plants outside the confines of their functionality and beyond the reductive consideration of those with 'taste.' Marder's point that plants don't think or see in images, but they do think and see, also reminds us of how art can create an experience that shows us something we couldn't see or understand before; art creates a level playing field for everyone to experience sensory stimuli. This equality is in keeping with vegetal experience.

Ascribing greater status to plant life is a political act, and art that draws attention to this shift in structures of legitimacy is also political. Jacques Ranciere's aesthetic regime focuses on the political capacities and sensations of art. Art is no longer a self-purge, he says, but a moment in political time.[7] The way in which an ecosystem flourishes, the way its elements vie for and share nutrients or don't, is political. Where Ranciere moves away from conventions of aesthetics and concentrates on where and how an artwork or art experience exists in a particular moment in time can also be connected with critical plant studies in terms of time. Like Ranciere's regime of political time, the elements within a given plant environment exist at a certain time and a certain place. The force of these extraneous constituent elements are a part of the work itself. Plant life, too, exists as part of a wider ecology and also as discrete individual ecologies.

In keeping with Ranciere's political regime for aesthetics, this book places art and plants on the political environmental agenda as an inter-species interrelation. It also highlights the necessary social and moral functions of creating a new aesthetic that relies emphatically on plant life. The artists discussed in this volume have been selected due to their synchronized connection to the critical plant concepts that have been developed and include artists from across the world, whose work I have come across during my inquiries to date. An online database is being developed to gather data on all artists working in this area and it will be maintained as an ongoing adjunct resource.

---

7 Jacques Ranciere, *Aesthesis: Scenes from the Aesthetic Regime of Art*, London: Verso, 2013, 11.

The Plant Aesthetics

Environmental aesthetics is an emerging field, as is Critical Plant Studies. These are innovative, if not radical, areas of research due to the sense of impending climate crisis that many of us feel we are living in. The emergent scientific evidence to support what we already suspected – that plants have capacities and capabilities beyond human comprehension – has been a monumental impetus for plant scientists, philosophers and art writers. Plant science (care of, Gagliano, Simard, Chamovitz et al.) shows us that plants communicate chemically, that they make decisions that suggest associative learning[8] and that they communicate using complex fungal systems amongst their roots.[9] Plants sense danger and vibrations, they manipulate insects and animals in order to thrive. We now know that plants remember.[10] They can function asexually and usually prosper as a community.

These are plant qualities many humans in Western culture have forgotten, or have lost touch with. Art and narrative writing has the ability to tie together plant knowledge and cultural studies, so that we can re-learn why our connection with plant life is so important. It is also important so that humans can see that, despite all the watering, fertilizing and positioning of plants that we do in our gardens or balconies so that plants grow, plants do not rely on humans for anything (except perhaps as composting, once we die). We breathe out useful carbon dioxide but so do myriad other animals. The philosophical and economic relationship between humans and plants has been patronising at best, and life-threatening at worst.

The discourse of plants and art falls within enviro-aesthetics. All that we yearn for, within a context of aesthetics, stems from conscious spirited awareness, a sentient life. These are not attributes we have historically associated with plants. However, the discoveries in plant science over the last decade have led us to reconsider the convention of denying plants a sentient life. If aesthetics creates a collective human mediation of the world, for a better understanding of being, then the varieties of plants, with their decidedly non-human and

---

8   Prudence Gibson, 'Pavlov's Plants,' *The Conversation*, 6 December 2016. https://theconversation.com/pavlovs-plants-new-study-shows-plants-can-learn-from-experience-69794. Accessed 6 January 2017.

9   Suzanne Simard, 'Leaf Litter, Expert Q and A,' *Biohabitats*, Vol 15, Edition 4, 2016, http://www.biohabitats.com/newsletters/fungi/expert-qa-suzanne-simard/ Accessed 6 January 2017.

10   Monica Gagliano, 'Experience teaches plants to Learn Faster and Forget Slower in Environments Where it Matters,' *Oceologia*, Vol 175, 2014.

seemingly mechanical activities, do not at first seem to complement aesthetics. Yet, increasingly, artists and scientists are drawing attention to the similarities between plant life and human life, and between plant life and media technology. These art and plant-science intersections are the connections I hope to make. These are also the creative elaborations I want to develop.

The Unclassifiable

The epoch of plant discovery, from the turn of the 21st century onwards, marks a profound change in human perceptions of nature. This has resulted in a period of extreme confusion as artists, writers and other humanists grapple with the philosophical implications of science research that shows plants remember, learn, associate and decision-make. As an Australian, this difficulty excites an alarming memory of an earlier time. Australia was invaded by white Europeans around 1788, when German Idealism and the Enlightenment were directing scientists to observe and to categorise, to denominate and to taxonomise the examples of plant life collected on sea voyages. The aesthetic decisions and praxes of the time were firmly entrenched in Romantic conventions. That is, they were still referencing classical motifs and Picturesque structures of creating paintings.

Once landed on Australian soil, the unwieldy bushland and the unclassifiable characteristics of the bush, the native plants and their foreign forms were too much for colonial artists. European mimicry and artifice flourished in early Australian art, due to a kind of paralytic inability among artists to create new visions. They simply did not know what to make of these strange new flora and fauna, and they couldn't absorb the oddities of new information into their aesthetic vision. Whereas back in the 18th–19th century, Goethe, Humboldt, von Neumayer and other roving scientists were searching for a unified whole, a system of recording nature that could be harmonised by an archetype; scientists are now searching for information about plants that transcends the hierarchies we have adopted since Linnaus. In other words, German Idealists yearned to find a plant archetype. Now we, or at least I, yearn to find a distributed plant knowledge system. This is Critical Plant Studies: to identify how these disruptive times affect culture and society's relations with the plant world, and to document the philosophical and aesthetic fallout. Both periods of history, the Enlightenment and the Anthropocene, are bound to disrupt and inconvenience what we thought we knew and who we thought we were.

Therefore, I am arguing that the same problematics are presenting now. This book interrogates art being made in this framework of new Critical Plant

Studies. There is a need to negotiate the criteria of aesthetic value of current plant-based and plant-theoretical art work. There has not yet been enough critical aesthetic discussion of the artwork being made that focuses on plant studies, despite the magnitude of enthusiasm artists are experiencing. There is some aesthetic refinement and critical formulation to be done.[11]

If plants are performing their own subjectivities, as I will argue they are, the classical tradition of aesthetics is made difficult to start with. Art has a long history of object-based making. Even performance and video work relies upon the body as object, the experience as something which must be recorded or archived before we can make aesthetic sense of it. Much post-modern artworks challenged these object-oriented criteria for what constitutes art; but it is the post-human turn, with its focus on moving beyond the limits of subject/object dyads, that best addresses the problems we face once we accept that plants are subjects, as well as objects. Plants perform subjectivities and art does the same. This book is dedicated to illuminating how this occurs.

The Science and the Theory

Subjectivity is important when thinking of plant scientist Monica Gagliano's careful experiments with the Mimosa plant. As part of her plant memory experiment, she built an apparatus in the lab that could hold a pot plant and then suddenly drop the plant, safely within the apparatus. The 'dropping' initially caused the plants to respond by curling up or retracting their leaves. But this, the anti-plant neurobiology critics cry, is no more than habituated mechanism! Plants grow towards the light, their roots crawl towards water and nutrient supplies in the soil.

However, Gagliano's experiment went two steps further. Once she had confirmed that plants react to this sudden dropping of them from a height, she performed the dropping again and again. Over time, they stopped exhibiting defensive strategies of curling their leaves. This was not just reactive or habituated. This proved a degree of learning. Then she waited a period of a month and repeated the dropping. Again, they did not curl up or retract. In other words, they had not only learned but they had remembered.

---

11   There is great work in this field of art an plants by writers Natasha Myers and Giovanni
        Aloi: Natasha Myers, 'From Edenic Apocalypse to Gardens Against Eden,' *Infrastructure,
        Environment, and Life in the Anthropocene*, Duke University Press, forthcoming 2017;
        Giovanni Aloi (ed). 'Beyond Morphology,' *Antennae*, Issue 18, Autumn 2011.

Plant theorist Jeffrey Nealon notes that, for Derrida, survival is the uncon-ditional condition of living.[12] Change and adaptation, then, do not necessarily constitute desires or interests, but survival. Plants exhibit a drive to survive. The interesting difference, as Nealon suggests, is that this is not an individual drive but a participation within a collective drive. This has implications for nature as a whole.

Nealon's suggestion that plants function as parts of a whole is not 'vegetal indifference,' but vegetal co-existence. Marder's vegetal indifference refers to humanity's disturbing habits of apathy towards plants when they seem not able to contribute to their own existential being. In other words, the suggestion is that because plants don't have 'desires' within the construct of language, they are not as relevant as humans. However, what if plants have a unique set of existential processes, from which we are excluded? Humans have been so busy excluding plants from the human realm of sentience that we may have missed some crucial scientific knowledge from which we are excluded.

These relations between plants and environment, and plants and humans, are entangled and material.[13] The lexicon in which to discuss these new dis-coveries is as demanding as the experiments themselves. Can we use terms such as neurobiology and plant intelligence?[14] Gagliano's recent Pavlov's Peas experiments, where she experimented to prove that plants have associated learning abilities, show us that plants are able to respond to artificial cues (in this case, wind) rather than real cues (in this case, light). To associate wind with the reward of light, when there is no light, debunks our previous assumption that plants' response to light is an automatic response and non-cognitive. Instead, these experiments suggest that plants are making decisions.[15]

12   Jeffery Nealon, *Plant Theory: Biopower and Vegetable Life*, Stanford, Stanford University Press, 2016, 57.

13   Marijke Van Der Veen, 'The Materiality of Plants: plant-people entanglements,' *World Archaeology*, Vol 46 No 5, 2014, 799–812.

14   P. Struik, X. Yin, H. Meinke, 'Perspective Plant Neurobiology and green Plant Intelligence: Science, Metaphors and Nonsense,' *Journal of the Science of Food and Agriculture*, Vol 88, 2008, 363–370; Anthony Trewavas, 'Response to Alip et al: Plant Neurobiology – all metaphors have value,' *Trends in Plant Science*, Vol 12, No 6, 2007.

15   Prudence Gibson, 'Pavlov's Plants,' *The Conversation*, 6 December 2016. https://theconversation.com/pavlovs-plants-new-study-shows-plants-can-learn-from -experience-69794. Accessed 6 January 2017.

The Glasshouse

Excavating terms such as plant intelligence in science may require an equally radical approach in the humanities (including art and writing). To move across disciplines requires order and argument, but in this book there is also an effort to incorporate experience and memory, just as Marder did in his recent collaborative book *The Chernobyl Herbarium*. The benefit of allowing narrative to creep into the crossover of art and science is that it allows generosity and growth; it also concedes the chance of failure in any experimentation.

I have a strong memory of my grandfather's glasshouse. The front gate to his house, for example, was not much more than knee-high and it opened onto a liver-brick pathway leading down the garden. The house had two wings, like double sentries, and lead-lined window panes in dappled glass. It was a big cold house in winter, due to the shadow of the hill behind, but cool and dim in summer.

His glasshouse was located around the side of the house and along the eastern perimeter of the rear garden. A glass structure of five metres by four metres, it had the smell of manky earth and seaweed. The seaweed concentrate was used to fertilise the many pots of orchids along the benches. Above the benches were two long suspended water reticulation systems that emitted a haze of water twice a day on a timer. It was warm in this space, the glass was smeary with dirt and cast strange shadows across the orchids. There were two long benches on either side.

He fed the orchids the seaweed solution via a squirt bottle. *Not too much*, he used to warn, handing the plastic bottle to me. *Too much and it will raise their hormone levels and they will develop ugly extrusions on their stems*. I squirted sparingly, drinking in the aroma of fetid and manky salt and wet earth. *More than that*, he directed, *too little and they won't have enough nutrients*. This time in the glasshouse was precious to me but I soon noticed that my grandfather rarely cared for the orchids, he instructed his wife and various female offspring to care for the flowers instead. Whilst this memory of him showing me how to care for his orchids holds happiness, it also heralds a warning. How can women, like plants, move beyond our instrumentality? How can we disrupt old habits so that women have the space to grow and bloom without being assimilated into the 'other's' being? This question dances across the lines of this book, and is ongoing and irreducible.

There are several female writers who are holding the flag of feminist plant studies, such as Luce Irigaray and Elaine Miller. Elaine Miller's investigation of the vegetative soul, through continental philosophy, reminds us of Heidegger's term the 'groundless ground,' which displaces the fiction of a univocal origin.[16] No plant is the same, no plant *umvelt* is the same. Miller believes Irigaray's trope of efflorescence 'explicitly performs what we have called a plant-like reading.'[17] Miller explains that Irigaray's efflorescence complicates the idea of a single reduced subject and a singular phallic reading, whilst reclaiming the ivy-like growth.[18] In Deleuze and Guattari's introductory rhizome chapter of *A Thousand Plateaus* they talk about the One (the reflection of nature) becoming two.[19] Whilst their logic extends to computer and IT studies, ones and twos, their ideas also allow for the distribution and multiplicity that we are increasingly experiencing in our digital lives. For art, that means a shift from one-to-one (viewer to painting) to one-to-many (viewers and experience). However, the impact of art remains the same: localized, personalized and intimate.

Plant Writing

Is this writing that I am doing here today a kind of pharmakon? Nearly a poison, nearly a cure? Do we need a changed perception of nature to redress the relationship between humans and the plant world in order to lessen the effects of climate change? People will suffer eco-crisis fatigue if we keep hammering away at the same doom-and-gloom points. However, if we create something, a means of communicating that may initially lurk in the shadows of a creative approach to writing, then this might be a way for creative writing to become real – this is hyperstition. Create the narratives of bio-rights and eventually the laws will change too. This was the kind of experiment that Derrida conducted in his writing. Write it down. It may be a trick, it may be a curse/cure. It may not be reliable or trustworthy, but once it is deconstructed and rewritten, it is done.

Marder said, of plant writing,

> But, no doubt, more needs to be done, boldly and experimentally, to invent a way of writing that would respond and correspond to plant life. Patience plays an important role here, as does the absolute openness to the other. Connected to this, I always wonder how to give my writing back

---

16   Elaine Miller, *The Vegetative Soul: From Philosophy of Nature to Subjectivity in the Feminine*, Albany: SUNY Press, 2002, 182.

17   Ibid, 183.

18   Ibid, 183.

19   Gilles Deleuze and Felix Guattari, 'Rhizome' in *One Thousand Plateaus: Capitalism and Schizophrenia*, Minneapolis: University of Minnesota Press, 1987, 21.

to plants. My dream for Plant-Thinking was to embed seeds into its covers and to urge readers to bury the book after it has been read, letting it decompose and germinate.[20]

In this book, I adopt Marder's call for bold and experimental writing. I present vignettes and case studies of artists and scientists working with plant science. This is a means of making the invisible visible. Plants are inseparable from their environment. That is one reason why they have been considered as having a non-identity.[21] Humans think of themselves as independent, thoughtful and wilful entities. They see plants as trapped in their places. They are bound to the ground by their roots, unable to move far or travel across the planet's surface. This suggestion of immobility surely is the strangest critique of plants. In an effort to reduce them to a less relevant species than humankind, we have cast them as insentient and immobile, neither of which is correct (seeds travel in the wind, roots dig for miles and miles, vines and ground-cover can extend across vast distances).

Marder walks through a philosophical arbor of plant-related concepts in his writing. His theses move the prominence of vegetal life away from categorization and away from a solely human subjectivity. He speaks of plants as the fifth element: air, fire, water, earth ... and plants.[22] In this book, I would like to respond to Marder's work, by asking why have we not paid more attention to the high-functioning, communication and sensory intelligence of plants in previous decades, particularly in art aesthetics, and whether that emergent art/plant area is effective? A consequential question will be what effect a radical reinterpretation of vegetal relevance will have on art and culture.

In an effort to align the philosophical ideas of aesthetics and ontology with plants, Marder charts various continental thinkers who refer to plants in their writing, for instance, Nietsche's interest in vegetal digestion, Hegel's concession that the act of devouring extends to the nutritive properties of plants, and Freud's repression as an interference with flowering as sexual maturation and human ripeness.[23] When Marder draws a link between 'spirit consciousness' as being only partly exposed to the light, as are plants whose roots are

---

20   Prudence Gibson (ed). 'Interview with Michael Marder' in *The Covert Plant*, Santa Barbara: Punctum Books, 2017.

21   Ibid, 162.

22   Ibid, 6.

23   Ibid, 172–175.

hidden in the dark soil, we see his wish, the urge, the desire to connect plants with 'life.'[24]

The art world is a discipline that can provide an interpretation of the changed way we understand or conceive of plants. For instance, much of the artwork being conducted around plant neurobiology is performance art, such as Australian Cat Jones and Latvian Spela Petric, and is known for its durational extremes and temporal qualities. The temporality of plant life, likewise, functions on a different scale and under a different model from the temporality of human life. Although the sun, moon and seasons have a direct effect on the opening and closing of flowers and on their ephemeral ways, short plant lifetimes and tree longevity mean that the operations are different. Plants operate on time scales of growth, dispersal and regeneration that are very different from human experiences. Plants operate within time scales we can't understand. Consequently, artists who create technological interfaces that mediate or alter plant time, and mediate their cellular communication and metabolic activity, are especially interesting as a means of learning from plants. Mancuso in his new book *Brilliant Green* asks 'What does art tell us about the relationship between human beings and the plant world?'[25] The groundswell of plant-related artworks marks a time in history where we are turning to plants for better models of living and creative solutions to climate change problems, presented as aesthetic acts.

Razing Terra Mater

An additional resounding point of curiosity at the plant-science and art nexus is the role of the female. This discourse is part of the emergent state of radical reinterpretation of vegetal life. Nature, of course, has conventionally been cast as a womanly figure. A mother, a fecund vessel within which life can grow. Plants, since the 19th century burgeoning of female interest in botanical studies, have attracted women. This is a domain of thought and scholarly engagement that has a strong female heritage. Many male philosophers have written about plants and trees in their work; for instance, Marder writes essays on Aristotle's wheat, Leibniz's blades of grass and Derrida's sunflowers in his book *The Philosopher's Plant: An Intellectual Herbarium*. Marder follows ten male philosophers, but he only refers to one female: Luce Irigaray. Marder and Irigaray work together on different writing texts, both scholarly and journalistic. There is the

---

24   Ibid, 173.
25   Stefano Mancuso, *Brilliant Green*, Bologna: Island Press, 2013, 11.

sense they conspired, for Marder's intellectual herbarium book, to draw connections with her favourite plant, the water lily. However, there have already been deep connections with plants for Irigaray. She has said to Marder in private correspondence that, 'All my work develops as a plant grows' and she has said that thought needs 'to be ready to listen to nature, to the sensible.'[26]

Sex doesn't correspond to sexism in the plant world, a mode that humanity might learn from. Most plants have a male and a female part in the one plant. Some mosses have male plants and female plants. Some conifers have two types of cones; one is the stamen cone, the second catches the pollen if the wind is howling right. Flowers have both stigma and stamen in the one plant. These self-fulfilling processes of reproduction have enormous significance for a culture where gender and sexual politics are a constant source of change and fluidity. There are potential provocations for a re-thinking of the world, when individual species have hybrid sexuality. This book will address plant and art hybrids as provocateurs in an art-science discourse.

We can't escape the bind of only understanding the state of being through the sieve of humanness. This is why, according to Claire Colebrook, 'becoming-woman' is still required.[27] An alternative can be to shake the tree and find a women's movement that is pluralised amongst male and female and all other gender combinations, to create new political and cultural units of thought. Interesting, here, to reiterate that Luce Irigaray's choice of plant as a means of elucidating philosophical ideas in Marder's book is the water lily. The water lily is both rooted and it floats across the water. It moves on a fluidly female surface and the lily reproduces asexually with the help of insects carrying seed from the anther to the stigma. The water lily, then, relies on no man, will not be tied down and requires the help of an entire ecological community to thrive. This plant is the perfect analogy for my aesthetic approach to plant studies.

It is time to reject terms such as Mother Nature or Terra Mater as terms that paralyse women into a role of carer, nurturer and healer. Instead, it is time to embrace the plant as a feminine symbol – growth and transformation, a female subjectivity. As Miller says, 'Woman is supposed to have an essence that defines her as a woman, once and for all. She is relegated to the status of nature or matter, and in this sense can do no more that assist or ground man in the actualization of his subjectivity.'[28] In this book, I discuss artworks as part of Irigaray's efflorescence, as a blooming metamorphosis and as an endless

---

26   Luce Irigaray, *I Love to You*, New York: Routledge, 1996, 139.

27   Claire Colebrook, 158.

28   Elaine Miller, 'The Vegetative Soul: From Philosophy of Nature to Subjectivity in the Feminine,' Albany: SUNY Press, 2002, 188.

individuation. These bloomings cannot be confined or bound in discourse. Irigaray's nonhuman and non-animal feminine subjectivity is an always sexed subject. Elaine Miller notes Irigaray's thoughts when she,

> Proposes a feminine model of subjectivity, one that returns to a close connection to the philosophy of nature, and in particular to the figure of the plant. In doing so Irigaray is not suggesting that a return to unmediated nature – in itself an impossible task – would bring about a meaningful change for women. Rather, she implies that a return to and a reworking of the symbolics of nature might be a place from within the social or symbolic order from which to retroactively restructure the ways in which women's embodiment, natural role, and passage into subjectivity are thought, and thereby to effect a real change for women in the cultural order.[29]

In my own modest way and as an homage to Irigaray, this book's focus is also not a return to unmediated nature. It is, contrarily, bound in aesthetic mediations of plants. The artworks discussed in this book are a reworking of the symbolics of nature and are real examples of restructuring women's embodiment and passage into subjectivity, leaving room for sexual difference.

The Contracts

As mentioned earlier, to establish a plant contract, I first need to turn to Rousseau's *Social Contract* and then move to Serres' *The Natural Contract*. Rousseau's earnest call, to mark a difference between the general will and the collective individual will, was a proposal that the individual give up his or her sovereignty in return for the care of the state: 'Let us take it that men have reached the point at which the obstacles to their survival in the state of nature overpower each individual's resources for maintaining himself in that state. So this primitive condition can't go on; the human race will perish unless it changes its manner of existence.'[30] Research in plant science shows that plants share information and all plants, in the given ecology, benefit from the

---

29   Elaine Miller, *The Vegetative Soul: From Philosophy of Nature to Subjectivity in the Feminine*, Albany: SUNY Press, 2002, 189.

30   Jean Jacques Rousseau, '6 The Social Compact,' in *The Social Contract*, 1762. https://www .ucc.ie/archive/hdsp/Rousseau_contrat-social.pdf Accessed 6 January 2017.

resources that are communicated and also protected by that information.[31] Rousseau's acknowledgement of the individuals and the collective are important in terms of perception of the parts and the whole of any given group.

Rousseau said, 'But, besides the public person, we have to consider the private persons composing it [the social contract], whose life and liberty are naturally independent of it. We are bound then to distinguish clearly between the respective rights of the citizens and the Sovereign, and between the duties the former have to fulfil as subjects, and the natural rights they should enjoy as men.'[32] This blueprint for a civil society, of the community as one being cared for by 'the one,' was established only 250 years ago. In the history of Western thinking, this is not so long ago. An animal contract and a plant contract are yet to be devised, but the sentiments of Rousseau's pre-democratic collective of individual wills are interesting to note in terms of a move towards a plant contract.

The Natural Contract

Michel Serres elaborated a 'natural contract' and developed his idea from Rousseau's 'social contract.' These precedents are a useful consecutive model for analyzing artworks, and collaborating with artists in their 'plant' endeavours.[33] We are no more than renters on the planet and we must therefore be aware there is a bond to pay if we leave a mess at the end of the lease. Our worries these days, according to Serres,[34] are weather patterns and time. Climate change has affected our relationship to the weather, in the short term rather than the long term. How does that affect the long-term now – decision-making with longevity? He writes, 'But more than that is at stake: the necessity to revise and even re-sign the primitive social contract. This unites us for better and for worse, along the first diagonal, without the world. Now that we know how to join forces in the face of danger, we must envisage, along the other diagonal, a new pact to sign with the world: the natural contract.'[35]

---

31   Suzanne Simard, 'Leaf Litter, Expert Q and A,' *Biohabitats*, Vol 15, Edition 4, 2016.http://www.biohabitats.com/newsletters/fungi/expert-qa-suzanne-simard/ Accessed 6 January 2017.

32   Jean Jacques Rousseau, *The Social Contract*, 1762, 21. https://www.ucc.ie/archive/hdsp/Rousseau_contrat-social.pdf Accessed 6 January 2017.

33   Michel Serres, *The Natural Contract*, Ann Arbor: University of Michigan Press, 1995.

34   Ibid, 27.

35   Ibid, 15.

So even while some of the criteria for Darwinian and Linnaen classification and ordering of various natural life science species were a way of placing humans in the apical position, as species that can speak and move and think, there is new evidence that plants can do many of the same things as humans. Michel Serres writes, 'Would a new Eden emerge if we agreed to a Natural Contract?'[36] He is referring to Eve in the garden, where her toes grow longer in the viscous humus and she is a garland climbing around the great tree. There is a sense, in Serres' writing, that Eve is a plant, entwining and braiding around the limbs of the Tree of Knowledge, which was of course cut down by Adam to make the first dwelling. A plant contract seeks to make amends for that original error.

The Plant Contract

The Plant Contract is a promise to hold a candle for Michel Serres' ideas within the natural contract. A contract suggests that something must be given up in order to reach an agreement. I will give you this, if you give me that. I will stop doing this, if you stop doing that. The plant contract reminds us that we have forfeited our responsibility, within the pact, to the vegetal world. In doing so, we have lost something ourselves. The code of conduct, the unspoken courtesy of the high seas has been washed away. This plant contract seeks to renew the agreement. The plant contract tells us that human sovereignty over nature has been occurring for too long. The plant contract acknowledges that dominion over nature must stop and that we can each sign our own small contract to address the Anthropocene, which refers to the damages done to the natural world at human hands. This is a series of plant contracts on a small and individual, human scale. It is written in the spirit of the 'one by one.'

My personal tree spirit is the poplar tree. Its leaves flicker in the sunlight and represent irresolution. Despite owning up to being 'indecisive' as a personality trait, I can confirm that the poplar tree propagates using a rhizome. These underground rhizomes or creeping rootstalks can survive for thousands of years, even when the surface trunks and foliage have been devastated by fire or drought, insect attacks or fungus. The rhizome is a source of longevity, fecund with creative possibility. In asexual plant reproduction, the rhizome can act as a reproduction system. The tips of the underground or underwater roots can break off as new plants. This is one means of reproduction in the water lily,

---

36   Michel Serres, *Biogea*, Minneapolis: Univocal, 2012, 107.

couch grass and nettles. The attraction of the rhizomic model in plants is their subterfuge, their activity away from human eyes.

New discoveries about how plants feel and think seem outlandish, because botanical specimens have no human-like brains. Instead, of course, they have complex rhizomic systems and methods of communicating via chemical emissions or via the communicative power of fungi that grows among their roots. The electro-chemical activities caused by remembering previous stresses, such as animals overeating their leaves or extended periods of drought, are the same chemicals that humans emit when remembering stresses.[37] These may be procedural memories in plants, rather than emotional memories, but we have more in common with plants than previously thought.

It is important to understand our experience of the botanical world. There is a long tradition of cultivating plants: the activities of gardening, ordering, classifying, collecting, picking and harvesting. Not to mention genetically modifying them and using hazardous chemicals to enhance and augment their growth. Indigenous Australians used traditional methods of burning the bush for regeneration and would plant a tree if a community needed shade.[38] So it is human nature to admire a well-manicured garden, to enjoy the produce of an apple tree and to baulk at unsightly weeds. The question this investigation asks is whether we have been treading the wrong track of plant aesthetics. Better to consider the bounty and liveliness of plants in-themselves, rather than for-us. By 'for-us' I mean for our aesthetic delectation, for our sustenance, shade and oxygen. Can we instead consider plants in their own right, via the mediations of art?

The plant contract is a method of telling stories about plants and art together. It is an aesthetic contract and as such has systems, rather than rules. It harnesses aesthetics as an analogic act, a way of drawing connections between species and things. It has the courage to be playful and celebratory, even in the face of possible failures. This is what it means 'to essay,' to venture out and experiment even though the potential for success is slim.

Art usually functions as an open-ended inquiry of becoming and of change, without proscription. The discourse surrounding the artworks that align plant and aesthetics is not already made, already decided. It is presented only as provocations, just some water for your pot plant of thought. Miller summarises Irigaray's plant thinking by saying that 'Irigaray subverts traditional metaphors as a rhetorical weapon against the tradition that has worked to exclude and at

---

37   Daniel Chamovitz, *What a Plant Knows*, NY: Scientific American/Farrar, Straus and Giroux 2012, 131.

38   Bruce Pasco, *Dark Emu: Black Seeds*, Broome: Magabala Books, 2014.

the same time to assimilate women.'[39] I hope to embrace the traditions of metaphor, as Irigaray does, and to also allow them to blossom and change without
constantly being reduced.

The chapters of this book serve as clauses in a Plant Contract. They start
with the wasteland in Chapter 1. The iterations of the wasteland chapter are
not confined to the land but are the outskirts of all kinds of environments,
even those under the sea or within a bio-hazard zone. Artists who work with
the most resilient of plants and engender that suppleness in their own work,
are introduced here. Wastelands provide us with provocations about civilising
the wilds, creating class structures in terms of 'work' and upsetting the order of
how we see plants and their growing environments. Wastelands refer to spaces
that have had their usage changed as the result of toxic accident, corporate
misuse, civic repurposing and via artistic mediation. This chapter grapples
with definitions and outcomes of the wasteland.

The history of the Green Man chapter charts the foliage-spewing stone motif
that has decorated churches and cathedrals since the 2nd and 1st centuries AD.
Green Man, an architectural iconic motif, emerged seven centuries before the
number zero was used in mathematics and is a source of playful and apotropaic wonder. The foliate-faced, leaf-speaking Green Man is especially relevant
when considering our human limitations in 'speaking' about plant life: we are
bound by our ineffective vocabularies in this field of plant studies. This chapter
extends how culture/nature interactions in contemporary art now, as with historical Green Man motifs of the past, affect our desire to relate closely to plants.
Any human yearning to connect to the plant world in a hybrid mode collapses
the distinction between the self and the other. Therefore, this chapter also introduces the abject nature of human/plant hybridity.

Robotany is developed in the next chapter as an emergent order of post-
human interspecies connectivity. Extending beyond humanoids or plantoids,
this chapter charts the science of plant systems and the technological intersections of plants as networks and computer systems as plants. There is much
research into the circuitry possibility and root-probing functionality of plants,
mimicked in new technological devices, but artists are also flipping these concepts and engaging with ways to bring attention to the complex vascular and
communicative systems of plant life.

The legal complexities of a world surface when plants are believed to behave
in human-like cognitive ways. Nature rights. Although we are vilified for calling
it intelligence, the way plants function using distributed cognition creates a

---

39   Elaine Miller, *The Vegetative Soul: From Philosophy of Nature to Subjectivity in the Feminine*,
      Albany: SUNY Press, 2002, 195.

call for new legal and political approaches to earth or plant jurisprudence. The eco-punks in Chapter 4 present the artists, philosophers, poets and activists who create opportunities for environmental change. The dyad of ethics versus morality is extended to the law, which inspired the concept that a plant contract might be a means of proposing the right for plants' rights. There have been some epic examples of earth jurisprudence and nature rights globally but they are few. The greatest obstacle, apart from neo-liberal greed, to greater care of our natural tracts of land, is the separation of ethics and morals. This chapter elaborates work done in this legal mine field.

Marking the water lily as a feminist plant inveigles the cross-species and eco-feminist concepts by such thinkers as Plumwood and Irigaray and lays bare the differences between sexuate beings that humans repeatedly either ignore or reduce to a unified whole. The secret magic of plant properties has been utilised in dark magic and in Western anaesthetics and medicines too. The ungrounding of plants in the last chapter involves removing them from the environment they cannot be separated from. This disruption is temporary. Ungrounding is, in itself, a metaphor for the way we have mechanised plant life and the need to restore a synergistic relation with the vegetal world. This is not to suggest that we no longer include plant specimens in our medicines, our magic or our artwork but merely to take care when doing so. The geo-philosophy unearths the ugly and mournful last stratum of the earth's surface, but with a message of future multiple subjectivities, leaving behind the horrors of past exploitation.

Vegetal life and plant thought is an exhilarating topic, especially when grafted to aesthetics and these myriad examples of plant artwork. The chance to grow and effloresce, to allow the appearance of plants to show themselves to us rather than impose how they should look, is the thematic of the writing here. To write without a script, without confinement, without being enclosed by prescribed outcomes is invigorating. It feels like there is the chance to grow from here. That efflorescence is due to Irigaray's suggestion that vegetal and feminine openness toward the other, in themselves, is a way of giving service to life: 'Thinking is reunited with growth.'[40]

---

40   Michael Marder, *The Philosopher's Plant: An Intellectual Herbarium*. New York: Columbia University Press, 2014, 221–224.

CHAPTER 1

The Wasteland and the Wilding: The Aesthetic of Abandoned and Reclaimed Green Spaces

The wasteland is conventionally a space on the urban periphery which had a prior utility, but is now redundant. It is also a place that humans have abandoned, neglected or otherwise misused; a place that once had vegetal life. In this text, it is a site where one function was fulfilled and now there is potential for a new function. Whether it is a place where weeds sneak up through concrete or where there is 'always a dead tree,' whether it is a botanical or agricultural place that has been reclaimed or a reed and seaweed-heavy shoreline that has been repurposed, the wasteland exists outside order, outside enforcement and outside civic planning. Ownership is sometimes difficult to determine, and this is the reason it illustrates the fundamental qualities of a plant contract. Artists adopting and reimagining the wasteland space as an aesthetic for their art practice are introduced in this context.

A plant contract sits well with art work being made at these fringes. Such a contract proposes that by paying greater attention to plant life, we may see the benefits of returning to nature. Art has the capacity to shift our perceptions of the world. As a result of perceptual shifts, we may learn to forfeit some of our more devastating patterns of commercial human behavior, that is, all those associated with exacerbating climate change. Humans have no claim to own the land beneath our feet. We have no claim to hold dominion over nature. By remembering nature, after forgetting it for so long, we may see the errors of our commercial ways. Wastelands are evidence of those errors; they have a memory of former plant life and help to show us the apocalyptic future if we don't remember the plants. In the wasteland there are weeds, there are tree stumps. Crayweed wraps around old shark nets, there are vines overgrowing walls and fences. A wasteland marks the collision between nature and culture. It is in this nostalgic and melancholy zone that we can pause to remember what has been lost and to find new aesthetic values that carry a different kind of social weight. By this I mean there is pleasure in understanding past mistakes, there is potential in valuing the evidence of previous wrongs.

In his natural contract, Michel Serres speaks of how nature has been given up by humans, despite its benevolence in providing shelter, food and warmth. He urges us to notice that nature has lost its power, but humans have suffered an even greater loss. Humans now must face diminished natural resources and the consequences of climate change. The plant contract, in turn, proposes that we return to Serres' contractual model, where we agree to a silent pact, and where we forfeit some of our human indulgences. The natural contract was too vast to contend with, but the plant contract is re-scaled to a single plant and a single human. If we forfeit our sovereignty over the natural world, we may potentially avoid the annihilation of the human species. At the very least, the plant contract is an urging to start making ethical decisions regarding the natural world from this moment on.

A plant contract will never be binding, of course. A plant contract will not be signed by the plant world. It is a one-sided document. Serres acknowledged this problem in an essay on Virginia Wolf's lighthouse. There he admitted an error in his original natural contract, because to expect Mother Nature to sit at a table with humans and sign a nature contract was tantamount to animism.[1] This, however, is to forget the crucial point he made in his 1995 natural contract text, which was that we have only two choices: symbiosis with nature, or death. The wasteland is both a place of near-death and a place of symbiosis. It is alive but exists as a radical vegetal form. It is alive despite human meddling and its 'civilising.'

Michel Serres said, 'We are dying, in effect, of separating ourselves from the world; we will surely die of continuing to separate nature from culture.'[2] The plant contract, presented in this text, is an exhortation to Serres to continue his work, to continue to remind us of our pact with the world, with nature. This requires a forfeiting of human dominion over nature and it requires greater legal representation of plant life and it requires a better understanding of the ethics of plant care. In this chapter on the wasteland, it is a reminder of dangers lurking just up ahead. It must also be acknowledged that there is tragic beauty in the aesthetic of the wasteland that suggests the self-sabotaging nature of the human condition.

Wasted Beauty

These wasteland spaces have a mysterious allure due to their spontaneous changeability, their radical disorder and their defiance of resolute human constructs. The wasteland comes in many guises. If there were a sleeping-beauty

---

1   Serres, Michel. 'Faux et Signeux de Brume: Virginia Wolf's Lighthouse,' *SubStance* 37, 2, 116 (2008): 125.
2   Serres, Michel. 'Faux et Signeux de Brume: Virginia Wolf's Lighthouse,' *SubStance* 37, 2, 116 (2008): 129.

landscape of overgrown thorns, a wasteland might be a suspension of time. If there were a contrived and purposeful reorientation of a space, a wasteland could be an artificial place. If there were a toxic location where it's hard to imagine plants could survive or where there is no chance of survival, a wasteland could be a superhero mutant garden. If a community garden is grown on the edges of a decrepit plot, a wasteland is a social comment. The wasteland, then, is a series of imaginative and speculative spaces of resistance. Wastelands become a social statement, with the potential for cultural impact and with an ability to change perceptions.

The relevance of a wasteland discourse to a plant contract is its ability to be a controversial vegetal zone. Wastelands can generate anger over past operations that have been lost or misappropriated. They can create a sense of dread regarding what might be lost elsewhere in other seemingly purposeful locations. The wastelands are the result of human work consecutively followed by a lack of work. The wasteland is an extension of a deathly transition zone. It marks the end of one utility, whilst awaiting the next.

At what point, I would like to ask, can the wasteland be reclaimed or restored via art as a mediation? How do the plants in these wasteland spaces create an atmosphere of resilience and versatility? The artists in this chapter are reclaiming wasteland qualities. There is a parallel perceptive shift away from conventions of beauty and truth, and towards an aesthetic that reflects a bleak future that has its own attributes, its own limitations. Where Serres talks about the way we have become isolated from nature which has caused greater damage to the planet than any human v. human world war, we can see the traces of his ideas in the wasteland. The wasteland is an isolated and damaged place. It has the remembrance of nature and its instrumentalism but the terms and conditions of the space have altered. The natural contract has been breached.

Areas of wasteland have sometimes been thought of as disaffected places. Our own Australian desert landscapes were so frightening to early colonial settlers of the 19th century, that they were absent from historical and artistic records.[3] It wasn't until artist Sidney Nolan (1917–1992) painted his images of the bare and barren desert with drought-ridden dead trees and animal carcasses that we began to see Australia as a wasteland ahead of itself. This refers to the fact that indigenous people were forced off a country that had, for them, been benevolent, leaving the arid zone for colonial settlers who could make no sense of its hidden bounty. Instead, they forced English homes and

---

3 Australian art was notoriously Europeanised. See Robert Hughes, The Art of Australia, Melbourne: Penguin, 1966.

inappropriate gardens onto indigenous Country. The purpose and interaction between human and land had been misappropriated by white settlers, leaving tracts of land untended and abandoned, out of fear. At this time, Nolan was representing an internal emotional wasteland, the angst of Modernism, rather than a true expression of the desert but he was working in an era where there had been excessive land clearing, where there had been poor agricultural practice and industry was beginning to boom.[4] More contemporary critiques of Nolan's work argue that his vision of atrophy and dystopia were misleading and that deserts are bountiful and benevolent ecosystems. Either way, his vision was one of a wasteland, whether psychological or prophetic.

The wasteland and the desert are not mutually exclusive – at least not within the parameters of this chapter's discussion. In this text, a wasteland is an abandoned place, a space where a garden has been neglected, where a landscape has been misused and mistreated. Industry, resource development, agriculture, fracking and mining are the perpetrators of failed towns, disused mines, Chernobyl, deforestation, eroded landscapes and rivers that ignite with gassy fire. These latter wastelands have a habit of restoring themselves at least in

FIGURE 1   *Kris Verdonck. Exote 2011. Performance: plants, animals, fish etc.*
COPYRIGHT KRISTOF VRANCKEN.

---

4   Nancy Underhill, *Sidney Nolan*, Sydney: New South Books, 2015.

part, but they were formed at the hand of humankind since Industrialisation – the Anthropocene. Wild weeds sprout from brick faces, new growth slowly inhabits razed forestlands, flowers have grown up at Chernobyl defying the lingering radioactive damage.[5]

There is also a new beauty in the scene of an abandoned factory where moss has grown across a wall or jasmine has crept across the ground, leaving a carpet. There is always a familiarity to the strangeness of a wasteland. The vegetal emerges in the wasteland as a signifier of defiance and strength. The weeds and wild grasses are a resistance movement against the ordering and classifying of the human world.

The Big Plant Art Claim

If art is a social dynamic where people gather to see, to experience, to participate and to be changed, then art may have the capacity to change our attitude to plants, and nature more widely. There are many artists working with trees and plants at this juncture in time, and not all of them turn out as expected. Belgian artist and theatre maker Kris Verdonk created a strange artist-made ecosystem in his 2011 *Exote* (see figure 1), where species that usually threaten biodiversity were allowed to run rampant.[6] *Exote* was exhibited at Z33 House for Contemporary Art in Belgium and was subtitled 'an indoor garden for invasive alien species.' There were parrots, rabbits, bamboo, bullfrogs and more – all species that threaten biodiversity. These are plants and animals that thrive outside their natural areas. In *Exote*, Japanese knotweed was grown, which is edible with a similar taste to rhubarb and grows as tall as bamboo. It is strongly discouraged in Australia due to its ferociously invasive nature. It blocks drains and corrupts built environments.

To enter Verdonk's exhibition (invasive species) space, visitors had to don a white lab coat, gloves and white gumboots. The question the exhibition asked was whether the animals and plants within *Exote* were contagious or whether it is the human species that are the contamination threat. Rather than being a lush garden of Eden, dominant species spoilt the bounty by taking over the other threatening elements, leaving a wasteland of post-violence, of post-virus, a post-apocalypse. Thus chaos ensued and the outcome was a hideous zoo of overriding nature, causing all of the garden to be destroyed afterwards, thus creating an alternative wasteland. Wild bio-writing, hazardous experiments,

---

5   Michel Marder and Anais Tondeur, *The Chernobyl Herbarium*, Open Humanities Press, 2016.
6   Kris Verdonk, *Exote*, 2011, https://goo.gl/Q5H61c.

toxic devastations, stories of resuscitation: these can also be the kinds of narratives of the wasteland.[7]

Within this wasteland chapter of plant-art discourse, the criteria for artistic output might be seen as the artist's ability to change our perception of what/ where and how nature grows. To that I would add that there needs to be an artistic expression of how we understand the time scales (life cycles and growing rates) of non-human life and the chance to engender a change in the viewer so that they understand something new about how humans live in or near the natural world in the 21st century. These works are often developed at a point where there is a confusion between what we thought we understood as beauty, and how the curious and the inverted have an alternative beauty.

The Invisible Labour of the Weed

The strange 'beauty' of the wasteland aesthetic, then, is something we are coming to know as part of a continuous trajectory of post-industrial, media-drenched anthropocentric life. It is an emergent Critical Plant Studies phenomenon. Perhaps part of that new aesthetic appreciation of the wasteland is related to loss and lack. Language and sexuality creates something new in humans: for all that we lose, there is something to gain. The drive to strive, and the force of the will to survive and to continuously encounter others, is part of the natural condition of being human. But the cost is high.

Humans may never find the right terminology to talk about a place that has failed, a space that has been abandoned. We may not find the right vocabulary, the right perceptive dimension to relate to a world where no forests will roll into the distance, where no bushland will threaten humans with the possibility of getting lost, where none of us will have the luxury of admiring the surprising diversity of a desert dune. The space of the wasteland is an elusive one. It is disconnected from the objects in it and disconnected from any real identity, or at least anything that is identifiable. And yet humans recognise a wasteland.

There is a strange familiarity in the sneaking weeds and the falling down fences; the buckling concrete and the disused architecture. Italian-born and Sydney-based artist Diego Bonetto works in this in-between space between city and its outskirts, between the visible and the invisible. By that I mean that in what most people might see as invasive plant species climbing up a drain

---

7   Jevan Watkins Jones writes about being occupied with plants as part of a 'wild writing' course that is tied to the Bio Sciences: Susan Oliver, Blog, University of Essex, http://susanoliverweb .com/2014/06/09. Web page.

pipe and clogging the flow, or the roots of a massive flame tree busting up the nicely even concrete, Bonetto sees as new growth. These are instances of resistance to the artificial and of his particular kind of re-wilding.

Bonetto sees his weed-finding tours and his environmental 'identity walks' as a story-telling art practice. This falls into the category of socially engaged art work. Creative, participatory and community-focused, social practice functions as a critical intervention. Despite, or perhaps because of, the collaborative and informal methodologies of social practice art, there is a gentle (sometimes extending to fierce) activist mentality at its core. There is evangelical pressure by socially engaged artists on the audience to participate in the politics presented. This is a kind of respectful lobbying and Bonetto is presenting his doctrine for better living, for better perception and care of the environment that exists in urban areas.

The wasteland – in this case Bonetto's reclamation of a space – is not a case of the human reclaiming nature, but nature reclaiming the human. This reversal of the relation between human and nature is important because we are so often trying to tidy up nature, to yank out the weeds and to pretty everything up. In fact, what Bonetto really does is he draws our attention to how nature reclaims itself. He creates an awareness for the peri-urban environment, the wild parts of the city. Nature is creeping back.

Bonetto has provided people with self-guided tours and audio stories of plants growing spontaneously in unusual places. One of his projects consisted of a pocket of land behind the film screen of the Kingswood Drive-In in Western Sydney. As an abandoned garden, Bonetto undertook a botanical archaeology of the small landscape. He interrogated the memory of this landscape, once the humans had left the scene. This was a post-human study in the sense that he was interested in how other species connect with the human. In other words, I interpret his work not as anti-civilisation, not as anti-industry or anti-commercial even, but instead a study into the resilience of certain ecological sites.

Bonetto has a sensitive approach to this mode of study as he understands it as a cultural belonging, a sense of identity and history to a particular tract of land. As Bonetto says, 'our environmental identity is reflected in what is known to humans, even what it is exploited by humans. When cultures collect plant species, they recognise themselves through those species. If you think of Italy you think of rolling grape vines, of Australia, the gum trees and of Brazil the endless forest. The way we relate to the environment signifies who we are. To become one with an environment, the plants become a totem – the Scotch thistle, the Irish shamrock etc.'[8]

---

8   Interview with artist, by phone, 3 October 2016.

But what happens when that environment we have come to identify with changes? We must change with it. As the wasteland spaces that Bonetto explores become changed in terms of their usage, they re-emerge as new ecologies. They are marginal and marginalised places, affected by the Anthropocene just like rural places. New species grow up and take over the peri-urban places. They colonise and disrupt. We disregard weeds, but Bonetto notes they have an important role to play. Weeds are capable of coming to terms with a fast changing environment and need to be acknowledged for this skill. He says:

> How we engage with the environment is as top predators. We use narratives to engage with councils because we can't sustain the current situation – there are not enough resources. If we understand the environment then we foster a better appreciation of resources. We, as foragers, look after the land. Foragers understand that we must leave enough. We are foragers and we are stewards of the land I notice the damage of a high tide, I go out at 5 am and see the small patch of growth since my recent harvesting a week prior. Via metabolism, by spitting pips of berries we find, we are participating in the growth and dissemination of species. The seeds that stick to our socks are then moved to other places as we walk about. This is a mutual benefit contract. It is an unsigned contract.[9]

Although Bonetto's weeds cannot sign any mutually beneficial plant contract, he upholds his side of a plant contract pact. He has metaphorically signed a document, promising to honour the vegetal life he encounters with respect and care. His labour is invisible to many. Bonetto's work relates to the invisible labour of Rousseau's social contract. The people who work without pay, without acknowledgement, must be honoured. By drawing attention to Bonetto and his quiet political acts, we can see the work, we can hold his performative praxis with high regard. Timothy Morton says that weeds are located outside the parameters of agriculture and therefore fall into a 'Biosphere Which is Not One.'[10] This is a play on Luce Irigaray's book *This Sex Which is Not One*. However, Morton seems to conflate Irigaray's thoughtful ideas about language and being a woman, and a sexuate being, into a strange non-logic here when he says that it is perfectly logical to self-contradict and to then delve into the history of the paradox. This loses the original agency of Irigaray's points that women's bodies

---

9   Interview with artist, by phone, 3 October 2016.
10   Timothy Morton, 'This Biosphere Which is Not One: Towards Weird Essentialism,' *The Journal of the British Society for Phenomenology*, Vol 46, No 2, 2015, 5.

are neither one or two.[11] Weeds, of course, are not located outside agriculture but are intrinsic to it, as something that must be removed.

The Appearance of Plants – Pre and Post Anthropocene (Goethe's Legacy)

So much of our culture is tied up with imagery of forests and bushland, gardens and farming country. Films, stories, picture books and poems lavish greenery on our thirsty visual souls: stories extend from Enid Blyton's *The Magic Faraway Tree* to Tolkien's *Lord of the Rings* to Julia Leigh's *The Hunter* – set in the wilds of Tasmania, Australia. However, increasingly, the real image contradicts the mythology, as farmland is carved up and subdivided off. Swathes of forestland are devastated. Ancient, wise trees are cut down to build wider roads. In such a short period of white settlement, Australia must have devastated their land more quickly than any other continent. The Marxist concept of invisible labour and the commodified and fetishised object works in reverse here. We can see the processes of agricultural and industrial labour there, in the land with no trees.

The mythology of nature then has lost its magic. There is no fetish except for death. It's bleached, eroded, cut-open and razed. We can't really justify using metaphors of women, abundance and plenty to stand in for nature any more. Nature is not a mythologised 'other' that has super-human qualities. It is just as subject to damage and lack of care as any human species. If we thought of nature as the abundant 'other,' we now must see it as a patient needing intensive care. The damage wrought is all too visible.

You can see the scars when you drive past the open cut mines in prime rural land. You can watch the coal being carried away in snaking freight trains. Then, once all of the given mineral has been dug up, the mines are closed and abandoned, leaving wastelands. These wasteland spaces are especially curious because they are spaces that exist well away from the front-of-house, that is, from the way most properties or agricultural lands are presented. Again, contrary to objects of fetishised beauty, these places are not the glossy final product but they are the wounds which tell the story of the labour, the process. These are spaces where the activities are visible, but not considered to have the same grace or conventional beauty as the spaces made most suitable for a public viewing. These are the back-of-house places. These are the appearances of nature in our contemporary experience.

---

11   Luce Irigaray, *This Sex Which is Not One*. Ithaca: Cornell University Press, 1985, 29.

This is so removed from the pre-anthropocentric idealists of the 18th century. German poet, scientist and writer von Goethe embarked on his Italian journeys of 1786–88 and his experiences were recorded in his published diaries.[12] While he spent many pages discussing his drawing, the festivities of various cities, art in museums, clerical robes, various canals and the poorest of labourers working the hardest of all, he also became interested in the varieties and order of different plant species.[13] He says, 'What new joys and profitable experiences the southern regions of this country must have in store for me! It is the same with works of Nature as with works of art: so much has been written about them and yet anyone who sees them can arrange them in fresh patterns.'[14]

It is this concept of the appearance of patterns and arrangements that eventually stimulated Goethe to write *The Metamorphosis of Plants*,[15] in which he searched for an *urpflanze*, the archetypal unity of plant systems, where '... each leaf elaborated the last' (the poem).[16] Goethe, like Thomas Aquinas before him, believed in the hiddenness of plants. This hiddenness refers to the life of a plant that humans cannot access, the realm of another species that we cannot cross into. Such is our human hubris today, that we have forgotten how limited we are.

Goethe influenced the American transcendentalists, such as Thoreau, Emerson and Poe, particularly in his alignment of passion and romantic fervour, with scientific and first-hand observation. He had an interest in a fundamental and original form, from which everything is made. In *Metamorphosis*, Goethe wrote of gases absorbed into leaves and constantly notes how each part is like the other – for instance, petals, leaves, seeds – and how they relate to the sum. He writes of the eyes that are hidden under each leaf, such as no.85, which are those nodules of new growth.[17] Linnaeus' theory of anticipation, where a tree in a smaller pot bears fruit more quickly than a larger one, is flagged by Goethe. But Goethe also proclaims that Linnaeus' idea, that the inner and outer rings of the trunk bark gave the energy for blossom or fruit, is an error.

Goethe was a passionate plant lover whose ideas changed our perception of nature. Rather than seeing the landscape as a 'scene,' he embarked on a reclamation of the functions and complexities of plant activity, whilst always

---

12   Johann Wolfgang Von Goethe, *Italian Journey*, London: Penguin classics, 1962.
13   Ibid, p. 320.
14   Ibid, p. 171.
15   Johann Wolfgang Von Goethe, *The Metamorphosis of Plants*, London: MIT Press, 2009.
16   Ibid.
17   Ibid.

yearning for this ultimate basic form. This is interesting in a chapter on the wasteland because, while we yearn for the picturesque, the romantic view of a scene (we were raised with picture books and illustrated fairy tales that conditioned us to do so), we are also drawn towards the flipside of 'the perfect.' The wasteland is imperfection and it is the comforting familiarity of the broken, the destroyed and the disillusioned that appeals to the human eye as strongly as the perfect scene.

A contemporary artist who has responded directly to the plant theories of Goethe is art director and botanist Ursula Zajaczkowska who created a work, *Goethe's Ballet*. This interspecies ballet merges plant growth with dance. It presents a slightly comical male dancer emulating the growth of a plant in a lab. But it is also completely mesmerising. In Marder's interview[18] with me, he discusses excrescence as how plants appear. Zajackowska, a Polish botanist, responds to this idea of how growth of plants appear to us. She cultivated plants over a two-year period in her lab and asked her dancer, Patryk Walczak, to respond with his body to the way we perceive the effects of plant growth and movement on us. The dance takes place in a light-filled glasshouse-cum-laboratory.

Zajackowska says,

> What this video presents is a series of my thrills about indeterminacy in the world of plants. We have a lot in common with plants, however, our human perception is deceptive since it humanizes everything around us. There is a lot of haughtiness within this idea, because why on earth should a plant resemble a human being? Nevertheless, such perception may also be used to suggest that a leaf is an arm, an apex is a head, and that a plant 'bothers with' reorienting its body towards the sun. All the time, I find it hard to believe that the moves of these plants are real. All the plants from *Metamorphosis* are dead by now. The have been 'used,' so they can die. In scientific research, a lot of plants are killed without emotion. Still, I perpetuate some of them in photographs and microscopic slides.

Zajackowska is aware of the ethical problematics of experimenting on and with plants. She recently proved, in a peer reviewed paper, that the hairs of the petioles of the Cucurbita genus are reservoirs of hydrostatic pressure. She claims she was able to make, or at least understand, this discovery as the result of observations during the filming of the dance. Here, again, art is participating

---

18   Prudence Gibson, 'Interview with Michael Marder' in Prudence Gibson (co-ed), *The Covert Plant*, San Diego: Punctum Books, 2017.

in unexpected ways in the discourse and development of perceptions of plant life.[19]

Further Legacy of the Adventurer

This was an example of contemporary dance colliding with plant research, inspired by the writing and plant poetry of Goethe. This connects life before the Anthropocene with life afterwards. Another historical figure also influenced by Goethe and his writings of an *urpflanze* was von Humboldt (1769–1859) who adopted Goethe's idea of a unified nature. To put the historical timeline in perspective, Charles Darwin read Humboldt's *Personal Narrative of a Voyage to the Equinoctial Regions* just prior to going on his HMS Beagle trip. This was a time of war, revolution and repression, which often led humans to turn to nature. As we are now. Humboldt wrote sixteen volumes – 8,000 plant descriptions and 4,000 new plant discoveries, two volumes of zoology and anatomy, four volumes of astronomical and geophysical observations, and seven volumes of *The Narrative* works. In 1793 *Cosmos* was published, which was his magnus opus. Humboldt developed a theory that plants are zoned latitudinally, not longitudinally.[20] His illustrations meant it had wide appeal. His interests were tied up with politics and the pre-democratic *Social Contract* 1762 as expounded by Rousseau.

1799–1804 were important and defining years as Humboldt ventured out on an exploration with Aime Bonpland, where he developed a unitary vision of the world and its phenomenon. In the early 1790s Humboldt worked in mines, devising safety strategies for miners. He observed mosses, lichen and algae (nonvascular) and other subterranean vascular plants. All growing away from light. In 1793 he wrote a book on 'plants in the mines.' In 1795 Humboldt met Goethe (who had written *The Metamorphosis of Plants* in 1790) and was influenced by his 'organic morphology.' Both talked about the collision of art and science. By 1800 Humboldt and Bonpland were in Cuba. Then Colombia. They had access to famous plant biologist Jose Celestino Mutis' herbarium and his botanical garden in Bogota. Both became consumed by the idea of knowing patterns of growth at different altitudes. Humboldt was joined by Caldas, a South American botanist, on the trip. He was a local and shared his knowledge on the next journey leg through the Andes. Humboldt's influence on current thinking can be listed as: 1. Plant communities 2. Eco-diversity 3. Vegetational

---

19   http://thecreatorsproject.vice.com/en_uk/blog/plant-growth-ballet.
20   Wulf, Andrea. *The Invention of Nature*. London: John Murray, 2015.

inertia (resisting change or invasion) 4. Mutual influence of vegetation and environment.[21]

These historical figures, so curious about plant life, are important to a discussion of how to make sense of the vegetal world and how relevant it is to human "being." An artist who has a strong affinity with Mutis but who has a history of reclaiming the damaged and the lost is Sydney-based Colombian artist Maria Fernanda Cardoso. As a young girl, Cardoso's father gave her and her sister each a volume of Mutis' plant taxonomies. Too big and heavy to bring to Australia, this vast tome stayed in her family home of Bogota.[22] Still, the influence of this early observational scientist has left its mark on Cardoso.

Cardoso is known for incorporating taxidermy frogs, birds, emu feathers and eggs, skulls and bones in her work. Even as far back as 1989 Cardoso was working with plant species. She was consumed by the importance of corn as a staple diet and a national agricultural product in Central and South American countries. The significance of corn, for Cardoso, became a curiosity about cultural relations. So she enacted language, through vegetal elements. Using corn kernels, she planted seeds in the shape of alphabetical letters and watched them grow as high as her thigh. At Yale University, she also grew

FIGURE 2   *Maria Fernanda Cardoso. Pollen 2016. Resin, site installation.*

---

21   Andrea Wulf, *The Invention of Nature*, London: John Murray, 2015.
22   Interview with artist May 2016.

clumps of varieties of indigenous and artificial corns – the colour and vitality of these vegetal art forms were saturated with meaning outside language but inside nature. Finally, she wove together all the corn kernels into massive coiled objects.

'We can't help being anthropocentric. Perhaps it's our biology to always see the human species so highly [in terms of status]. But we can try to move away and see the natural world more closely. We can try.'[23] Cardoso certainly does look at the world closely. She has always dealt with the minutiae of nature. Using microscopes to train fleas for a *Flea Circus* 1994–2000 and high tech video equipment to film stick-insects swaying in the breeze has been her default performance work. She is interested in how the wind is invisible but we can see it by the way it affects the movement of trees and creatures, and ripples the surface of the water.

She has included plant elements in some of her large-scale insect works such as her 2013 *Museum of Copulatory Organs*. One of these elements was a cabinet full of cast sculptural pieces that mimicked pollen. From biology books, Cardoso created three-dimensional versions of botanical illustrations, coupled with Neolithic stone forms. These small works were sized to fit in the cup of your hand. Cardoso was interested in genealogy and Linnaeus' system of categorising all plant genus as sexualised. So, she examined the sexuality of pollen. Pollen is the male seed which reproduces with the stamen. This self-serving, self-generating process provides the morphology of identification.

Cardoso is most relevant to this chapter on wasteland because of her ability to reclaim the tragic history of her country – Colombia, a place that has been decimated by commerce, war and political strife. But also because of her interest in the minutiae of nature – the will to reclaim and herald the wonder of the microscopic in nature. The wasteland gives the immediate appearance of loss, of lack, of abandonment. But look closer and we see the resilience and restorative energy of so many disused sites. The word 'disused' is perfectly deceptive as it is only the human that has decided on a change in use. For the ecosystem that exists, life goes on.

Cardoso was commissioned to create a larger sandstone constellation of pollen artworks (see figure 2), destined for a Darling Harbour re-development on Sydney's foreshore in 2017. Darling Harbour: surely the ugliest of waste-lands! These larger works were in response to a report on pollen, undertaken at the Darling Harbour Lendlease site to gauge dust in the precinct. The report then fuelled Cardoso's artistic interpretation. The dust specialist found some

---

23   Ibid.

fossilized pollen during his investigation of the harbour-side area. Through this discovery of microfossils, a story of dust and pollen unfolded for the audience.

For her *Naked Flora* 2012–14 series, Cardoso stripped the flowers of their petals, to investigate the stamen (male) and pistil (female) reproductive parts. Then she photographed the flowers using macro-stacking – a method of taking many images to create a depth of field, a four dimensionality that turns the images into sculptural forms. Smaller flower body parts were displayed in cases. Her work is deeply influenced by the 1735 classification system of Karl Linnaeus. He researched the sexual systems of plants as a means of classification through ordering and measuring and cataloguing. Cardoso's images were tentacular, colour-saturated and alien. She had a workshop at the Parramatta Public Art Gallery, and encouraged participants to strap on a head magnifying glass and, after investigating the inner apparatus of flowers, to draw and record what they saw.

You have heard of the carnivorous plant, the Sarracenia, which has a tubular shape with a hood. Insects are lured to the nectar on the rim of the tube. Laced with narcotic, the nectar causes the insect to fall into the tube where they are digested by the enzymes before they regain consciousness. The effects of Cardoso's photographs don't result in this kind of finitude, that is, death. Instead, the finitude of her photographs subverts the way we think of plants, as the plants might instead think of us.

The Interior Wasteland

Zajaczkowska, who directed the Goethe ballet, Cardoso, whose obsession with plants was generated by Mutis, and Tega Brain (an Australian artist working in New York) have been included in this chapter for two reasons. First, they are responding to the growth and appearance of plant life in different contexts. Secondly, the reality of a wasteland is that it is not (any longer) limited to an outdoor locus. Zajackowska's direction of the dance video was in the white and muted light of a glasshouse that was also a laboratory wasteland. The background of the video was a cool and purposeful place where industry (science) had been undertaken. Experiments were done, gloves were worn, scientific equipment was used and the entire space was hygienic and clinical, and all waste disposed of appropriately. Yet, as the botanist explained, there was death in that lab. Her experiments with plants resulted in many deaths. This is the wasteland. The place of death.

Likewise, Tega Brain conducted her *Coin-operated Laundromat* (see figure 3) in an unlikely place. *The Laundromat Wetlands* emerged after previous works that involved plants, such as *Le Temps*, and *Keeping Time*. *Le Temps* explored people and plants: the phenology of seasonal patterns, and the buddings of plants, seasonality, responses to changes in climate. In *Keeping Time* 2002–2013, Flickr digital images were searched and compiled. Botanic species included the Sturt's Desert Pea, jacaranda, cowslip orchid, the Texas bluebonnet, etc. The photographic images were time stamped and then laid out, to see seasonal patterns that emerged. Some species get seasonal attention. She says,

> The work also shows patterns of species visibility. Observing the number of photos taken throughout each year. Plants only become visible to us at particular moments in their life cycles – typically when a species flowers, but also when leaves turn red during fall and bud in spring. As such this project provides glimpses of widely held socio-cultural relationships to the species in question, showing some to have cultural significance as people's names, names of places or use in events like festivals or weddings. The patterns in the resultant images are ambiguous, pointing to both seasonal and cultural correlations, deliberately resisting what theorist Benjamin Bratton refers to as the "staged transparency" of much visualization work.[24]

Her interest in phenology was piqued by climate change because it is a barometer for biospheric destabilization. The changing seasons and the movement of time places this well in a discussion of plant-human time schisms.

For her *Coin-operated wetland*, she created a wetland in a laundromat. Wetlands filter water, making it clean to drink and wash with. Brain was aware that the wetlands are often thought of as nasty smelly swamps – wet wastelands – and are constantly under threat of development and improvement. So she moved the small wetland into the safety of a laundromat. Brain raised the stakes by marrying two kinds of wastelands – the wetlands and the laundromat, where wet washing and dirty laundry adds to the aroma of the manky wetlands themselves. The artist, who has a background as an engineer and builds wetlands for the City of Sydney, believed the work collapsed the division between human and environment.[25] The wetland's reedy plants were to create energy to drive the washing machines of the laundromat.

Tega Brain says,

> I was interested in the health of the downstream environment. I wanted to set up the system as a way of showing how most infrastructure systems

---

24   http://tegabrain.com/Keeping-Time.
25   Runway issue 21 'Ripe.'

FIGURE 3   *Tega Brain. Coin Operated Laundromat 2011. Performance: washing machine,*
             wetland plants, pump, tank.

seem to have no negotiation. I wanted a new system to have negotiation
between wetland and human participant that then makes an assessment,
that is, to see how they behave. Engineering is about risk management,
like the inside/outside of an ecosystem. Environmental politics is running
out of systems. Projects are dysfunctional because water quality does not
comply.[26]

The Romantic Wasteland

As Matthew Gandy points out in his *Marginalia* paper, 19th century
philosophers such as John Ruskin saw nature as having a life force or being
outside the transcendent, that is, located in its materiality, via observations in
the field.[27] Back then, Gandy notes, Henri Bergson focussed on the materiality
of nature, his positivism stretching out to Deleuze and a material reality.
Deleuze's material reality moves away from Bergson's human finitude, to the
beginnings of post-finitude – a type of knowing that exists with or without

---

26   Author interview with artist by Skype 22 February 2014.
27   Matthew Gandy, 'Marginalia: Aesthetics, Ecology, and Urban Wastelands,' *Annals of the*
      Association of American Geographers, Vol 6, No 103, 2013: 1308.

humans. What does this mean for wild places? What does that mean for the in-between areas that link the structures of gardening and landscaping with built structures and neglected or abandoned industrial areas?

My answer is that it introduces the ecologically aesthetic concept of the wasteland as a place that struts its own beauty. The distaste we have for rubbish bins and manure piles, for offcuts of rubble and composting bins is intrinsic to the wasteland aesthetic. This is where green matter, old turf, branch cuttings and the wood chips of old dead trees are left behind. Repulsion is a contravening of visual and olfactory conventions of pleasure. However, as Edmund Burke explains in his 1757 *Philosophical Inquiry into the Beautiful and the Sublime*, repulsion functions as a dyad of experience. It is a counterpoint to pleasure.[28] What disgusts us also creates desire and pleasure, especially if the danger of exposure is at a distance. We can apply this idea of disgust to wastelands. Images or even physical experiences of wastelands can provoke pleasure and sensations of desire in a viewer, so long as the viewer can escape the experience shortly afterwards, or so long as the sight is experienced at a distance.

This is a romantic sublime notion. It is not quite aligned with the philosophical associations of Critical Plant Studies as developed in this book. Here, CPS is a perception of the various ecologies of the world as fractals of energetic activity. Each ecology connects with the next and to its entirety as well. This is an aggregated system of approaching changed notions of nature – away from transcendence, away from sublime aesthetics, away from the human mind thinking like a machine. Instead, the human mind approaches its environment as a plant system. Vegetal studies can help to understand human connections with its ecologies, and human cultural practice can help humans to assist plant functions in order to provide a more cooperative way of living. It is the fundamental place of plants in human lives, irrespective of the relevance placed upon them or not, that makes this subject area a tantalising segue into ecological variegations such as the wasteland.

A Bacchanalian Labour

If the plant contract is an effort to redress damages incurred upon nature due to human over-dominance and extreme sovereignty over the natural world, then Jo Burzynska is playing a role in adjusting our sensory understanding and

---

28   Edmund Burke, *A Philosophical Inquiry into the Origin of our ideas of the Sublime and the Beautiful*, London: Oxford Uni Press 2015, 1757.

awareness of nature, of wine vines in particular. She is experimenting with changing our perceptions of the vegetal life, and is dabbling with the changed usage of a given landscape. Burzynska is a UK-born and Sydney-based sound artist and has worked for many months on those very rolling grape vine hills of Italy that Diego Bonetto was discussing (when he mentioned how we collect plants to create identity). She has a strong track record in wine appreciation and is doing interesting work at the intersection of perceptive taste and sound. You can undertake an experiential example of work on her website. One of the works listed there is *Bittersweet*, a magical sound track that starts with the low drone of fermentation equipment, which soon ranges up to a higher pitch of cowbells, cicadas and distant birds, and then returns to the dark hum of wine manufacture. Burzynska's research, undertaken with a group of participants, showed that the taste of chocolate was experienced as more sweet when the sound frequency was at a higher pitch.

So I tried it out on my son. I fed him squares of dark chocolate and played the sound track from my computer, sitting right next to him on the couch. But, oh the failure. He experienced the opposite. For him, the chocolate tasted more bitter when the higher pitched sounds were played. 'George,' I cried. 'You got it back to front.' My son apologised, shrugged and turned back to *The Simpsons*. Okay, fine, I thought, he was just distracted by TV. So I tried it too, but to my horror, my experience was the same as my son's. The chocolate tasted sweeter with the low drone rather than the more ebullient and abundant sounds of 'nature.' Both my son and I failed the experiments, but perhaps the outcome doesn't matter as much as the change itself. The sound track did *change* our sense of the chocolate and that, fundamentally, is Burzynska's premise.

The artist has a background in cooking and wine-tasting, running a wine shop and getting her wine qualifications. In 2012, as part of an artist's residence in Campania Italy, she was able to witness how wine makers listen to the fermenting vats. Her idea was to make oenosthetic art. She recorded sounds of the vineyard and of the fermenting vats, the sounds of the training wires for the wine vines whistling in the wind to create a soundscape. Her plan was to knit together sound and taste so that it might change perception.

She says that she,

> became aware of the growing body of research in crossmodal psychology that suggested the senses could indeed influence each other, which included a small but intriguing number of studies directly concerned with correspondences between sound and taste. One of the first I encountered was Adrian North's study in 2008 using wine and music, which suggested that music is not only able to prime certain thoughts and feelings, but

could have an impact on the perception of taste. This was supported by subsequent studies that started to map elements such as pitch and timbre to basic tastes.[29]

All of this falls into the concept of the *terroir*, which is a French wine-making term referring to the entire environment – soil, topography, etc. – in which the wine is made. To the *terroir* that she was working with in Campania, she added sound. Her soundscape was the reverberations of nature and of wine production. She then set about playing her soundscape track to people and testing their changed perception of wine or chocolate. She found that most people had the same sensory experience and that the responses were clustered. The aqueous sounds of the sparkling wine created particularly heightened perceptions. What also happened was that the artist became quite adept at understanding the soil ... becoming an agronomist of sorts. She also noticed that much of the wine-making language was sonic – e.g. off-notes.

I am interested in interrogating different kinds of wastelands in this chapter which, by definition, is the movement from a used space to an abandoned space. These are physical and emotional spaces. To find the place that is no longer useful for humans, that are also psychological spaces. They offer a new way of being, a new way of perceiving (hearing and tasting). This is the force of Burzynska's work, that she is creating a useless (non-instrumental) experience immediately out of a useful one. From the production of wine, comes her experimentation with oenosthesia, which at this stage has no real applications. In some ways, this is like sowing new worlds. New germinal wisdoms, new ways of seeing hearing and tasting the world are borne out of wasteland spaces – out of uselessness itself.

These ideas are relevant to critical plant studies and the presentation of a plant contract because labour and activity with nature, rather than from its utility, is a crucial area of potential change. If a plant contract can be no more than a mode of respecting the land for what it is, then working alongside wasted spaces, and changed uses in industry are important components of consideration.

---

29   Burzynska cites the following reference: Crisinel, Anne-Sylvie. Spence, Charles. Cosser, Stefan. Petrie, James. King, Scott. Jones, Russ. A bittersweet symphony: Systematically modulating the taste of food by changing the sonic properties of the soundtrack playing in the background. *Food Quality and Preference*, April 2012, Vol 24(1), pp. 201–204.

The Wasteland of Nuclear Fallout

A wasteland is not limited to deserts or vineyards, drive-in movie cinemas or strange laboratory dances. There are places where the land is not fighting back, where it is not being resilient, such as eroded, salt-laden ex-mining sites. There are tracts of land where the beauty of a wasteland aesthetic is in evidence but the cruel sting of a tragic event makes it difficult, if not immoral, to bundle this site in with an abstract discussion of beauty or lack thereof. There is a place, a wasteland of toxic super powers, that could be the greatest inspiration for a comic book tragedy that becomes a superhero story.

Imagine being only eighty kilometres down the road from Chernobyl when the nuclear plant exploded. Philosopher Michael Marder was on holidays in Anapa, Southern Russia, on the Black Sea that day in 1986 when the graphite reactor exploded at the nuclear power plant. Over 350,000 people were evacuated from the surrounding Ukraine, Russian and Belarus areas. For thirty years, the site and environs have been not just a wasteland, but a ghost land. This area was not neglected and abandoned when no longer needed, like many other wastelands. This was an accident-derived wasted place. There is an element of fear and danger in this wasteland which sets it apart from others, and as such doesn't have the peaceful melancholy of Bonetto's drive-in cinema nor the implied bloodshed and political angst of Cardoso's earliest Colombian works.

Michael Marder teamed up with artist Anais Tondeur in 2016 to create a book that combined philosophy, memoir and photography. They aimed to reclaim the Chernobyl wasteland, to give photographic evidence for the versatile plants that have slowly grown up in this hazardous explosive space. There are white patterns, tiny little specks of light-dust that may be radiation, on Tondeur's images of the plants from the herbarium of the exclusion zone. They seem ethereal, even wasted, losing water or moulting at the wrong time.[30]

The terrible irony of Marder's visit to Anapa was that he was being taken to different climes because of his severe allergies to birch and oak pollen. As he explains, he had been living as a boy in an apartment block on the outskirts of Moscow, between a massive forest and a polluting industrial factory. So off he went on his train trip to a healthier place where he, in fact, would receive dangerous doses of radiation from Chernobyl. As Marder says, 'Jean Baudrillard dubs this the logic of seduction, of fleeing towards the thing we are trying to escape.'[31]

---

30   Marder, M. and Tondeur, A. *The Chernobyl Herbarium*, Open Humanities Press, 2016.
31   Ibid, 15.

Tondeur's photographs, such as her geranium, are specimens that are monitored and constantly assessed. Marder was interested in the process and quality of exposure, as allegorised in Tondeur's photos. These plants that have grown in radioactive soil have been exposed to radiation, and are also exposed to Tondeur's camera, her printing process, and inspired Marder's exposure of his own story. 'Vegetal life is not merely exposed,' according to Marder, 'it is exposure, exteriority, outwardness.'[32]

Human connections with nature (in all its cultivated and constructed iterations) are deep in our psyche and yet the vegetal world changes with more speed than our imaginations can keep pace with. So the plant philosopher and the artist have created this collaborative book of essays and photographs of the new life springing up at the Chernobyl site. It is a celebration of plant versatility and endurance, a visual and textual plant contract.

Strangely, a forest of pine trees at Chernobyl ground zero turned red after exposure and soon died, having fallen to the earth. The odd thing is that they are not decaying properly. The time scales have been interrupted and the progress of decomposure is out of kilter. Yet, out of these strange Chernobyl details, there is generation of growth and of ideas, and of the kernel of the book Marder and Tondeur have made.

As Chernobyl showed Marder, we live after the end of the world. The plants still grow, the red pine trees remain biotic (in the sense they still exist even though they are clinically dead) and Marder has survived to tell this story and also led many of us on a growth spurt of vegetal curiosity. This Chernobyl wasteland is a dynamic place. It is most certainly alive.

The Wasteland Workers

Alive, and working. There is a road that runs along the rear of a major horse racecourse in Sydney.[33] From this road, you can hear the whinnying of horses in their stables and can smell the equine dung and freshly cut hay ready for feed. These are the working areas of the racecourse. There is a hut where the blacksmith shoes the horses. On some evenings, you can see his light glowing behind a window and the clanging of his hammer on the metal. Right there, in the middle of the city. What really intrigues me are the piles of sand and abandoned piles of turf. All off-cuts and 'green' waste matter are piled up in this back area of the property. These piles of old sand, dead grass and cuttings

---

32   Ibid, p. 22.

33   Wanesy Road, Randwick.

from the roses over on the good side of the racecourse have slowly grown weeds. They have turned into strange hills of abandonment. No need to tidy these weeds, no need to 'pretty up' these places. But the sparrows swoop and dive, fidget and peck at the weeds for their tasty seeds. Occasionally a small dozer truck will drive over and add another pile of green waste to the piles. This is a dumping area but only for biotic matter. This is a wasteland.

Is the wasteland space, then, an issue of class? If these wasteland areas have their own unconventional beauty, not usually seen as 'tasteful' for a discerning viewing public, then does it come down to the divide between labouring workers and the indulgent racing audience? On one side of the racing ring, the posh folk sip champagne and tend their million-dollar Arab horses. On the other, there is the busy activity of unbeautified grounds and unsightly refuse. Which is more beautiful? The clipped rose bushes and colourful dresses? Or the clanging of the blacksmith, the barking of the groundsman's dog? In this chapter, the aesthetic beauty of the wasteland has become visible.

The wasteland emerges as a misfire, a disjuncture in the move from social contract to natural contract to plant contract. Somewhere, the morality has slipped through the mesh and become something else. Old standards have been abandoned, replaced by new ones. Plants don't attribute ethics or morality to themselves. There is no individual good, but only the good of the collective. Plants thrive together in clusters, they can grow symbiotically and they communicate information about their environment to one another. We know this due to recent plant science, but we also know it intuitively and anecdotally. We can see the way species survive better as part of a larger green conglomerate than when they grow in isolation.

Untilled

The wasteland is not just piles of unsightly dirt and abandoned mess of tiles. It is the acknowledgement and declaration of those elements as a wasteland that make it so. The wasteland as aesthetic was taken up by Pierre Huyge[34] in his Documenta 13 work *Untilled* 2012.[35] Out the back of the outdoor sculpture and installation space, his work was scattered amongst the utilities and mechanical equipment buildings. These were spaces not meant for exhibition audiences. Here, where all the leftovers, the scrapped and disused materials lay, Huyge

---

34 Amelia Barrikin, *Parallel Presents: Pierre Hugye*, New York: MIT Press 2015.
35 Pierre Huyge *Untilled*, Documenta, Berlin, 2012. https://www.youtube.com/watch?v=mEj Ey3RY370.

placed his female nude sculpture with a beehive on her head. It was where his white dog, painted pink, and his caretaker with pink hair wandered aimlessly. The piles of abandoned tiles and building refuse were intrinsic to the work. *Untilled* was a clever political play on the idea of land that has not been turned over and made ready for agriculture. It also referred to the future. 'Until' is a preposition or a conjunction and always introduces something further, something that happens next. A subjunctive. It addresses the next clause, the next phrase, the next epoch, the next spatio-temporal state of humankind.

The curious aspect to Huyge's work is the element of tangential relevance – the element of speculative possibilities of the wasteland aesthetic. What Huyge's work does is dismantle our concepts of beauty and order, and help us to consider a sense of aesthetic enjoyment that sits outside desire. These spaces are areas of need. We need to leave messy piles somewhere; we need to have workmen conducting their labour. But do we also need to bear witness to this labour, these places of utility? We need to make visible the labour of these places – where weeds grow up, where vines wind around branches, where unruly plants are not cut back, where grasses are not fertilised.

Natasha Myers and Carla Hustak have written about the affective ecologies of plant/insect encounters, which happen to be present in Huyge's work in the sculpture with beehive head. For instance, the Ophrys orchid, as Myers and Hustak explain, attracts pollinators via mimicry and gives them no nectar reward. These are 'asymmetrical encounters.'[36] This term is so apt for reflecting on Huyge's work. Another work at the Sydney Opera House, *A Forest of Lines*, was part of the Biennale of Sydney 2008, and comprised thousands of pot plants borrowed from nurseries and brought into the concert hall. A light mist fogged the space and music was played to the plants. In contrast, in the *Untilled* work, the experience of the wasteland areas of the site made up the sum of the work: they were 'asymmetrical encounters.' Not uncanny, nor perverse. There were no contradictions or paradoxes. It was instead a celebration of asymmetry, of revelling in the irregular. Unlike Goethe and his search for the *urpflanze*, that perfect archetypal form, there is a place, especially in the wasteland, for the imperfect form. Myers speaks of involutionary modes of attention, as opposed to evolutionary logics.[37] This suggests the inverse of constant change and constant improvement.

---

36   Myers, Natasha and Hustak, Carla. Involuntary Momentum: Affective Ecologies ad the Sciences of Plant/Insect Encounters, *A Journal of Feminist Cultural Studies*, Vol 23, No 3, 2012, p. 76.

37   Ibid, p. 81.

The wasteland is that inverse mode of improvement. It is a celebration of all that is untended, left alone. Wastelands are where new elements in an ecosystem are entangled and relational but do not necessarily augment human lives. Myers' reading is a 'feminist mode as affective encounters,' referring to Elaine Miller on this point, which have been the privilege of the animal world to date. Another term for wasteland is 'brownfield,' which refers to a type of disused space that had pre-existing industrial uses on the urban fringe.[38]

The Social Pact

Although the plant world does not seem to adhere to a structure where there is a sovereign power, there is no way that the limitations of being human allow us access to whether this is correct or not. The 'secret life of plants' is more than a title of a bestselling book,[39] it is also a valid reminder that we still know very little about other non-human species. We are even incapable of making sense of weeds.

My favourite weed is Patersons' Curse, a fantastic purple flower that takes over inland Australian pastures in winter and ruins grazing land. This weed, so beautiful to see as it spills over a rural hill, is a terrible burden for farmers as it poisons livestock and uses up the nutrients in the soil, making it difficult to eradicate and re-plant a crop afterwards. It causes allergies and crowds out the native vegetation that is more palatable to livestock.

Now this idea, that Patersons' Curse takes over more nutritious and delicious vegetation that cows and sheep prefer to eat, is worth noting because it begs the question: is Paterson's Curse protecting itself and its familial native pastures from the chomping and pooing cow machines? Is 'the curse' securing its own (and its vegetal neighbour's) survival? Unfortunately, all we can do is speculate. We cannot access the language that plants share with one another, except to measure their chemical emissions. We cannot ascertain the intent or the desires of plants.

This dislike of weeds in rural locations does not really explain the disdain for weeds among urban gardens. So many weeds are beautiful. Think of the weeping willow, one of my favourite trees. Think of forget-me-nots and bluebells. Not to mention the much disliked lantana with its grab of tiny purple, pink and yellow flowers. Garden theory is control, management and

---

38   Gandy, Matthew. 'Marginalia: Aesthetics, Ecology, and Urban Wastelands,' *Annals of the Association of American Geographers*, Vol 6, No 103, 2013: 1302.

39   Tomkins, Peter. *The Secret Life of Plants*, Harper Collins, New York, 1989.

displacement of the wild. The wasteland is the reverse – a political enabling of weeds, wild growth and lack of authoritative controls.

Taking the High Line

Finally, this chapter winds up with a wasteland that is both contrived and made artificial. It is included here as the ultimate gimmicky invention, and yet it is important as a hybridised nature/culture environment. After all, it would be foolish to think that casting our eyes back to nature can be a purist process with only pristine past landscapes in our viewfinders. *The High Line* in New York City is an energetic dynamic revitalisation of a wasteland space. This nature/culture encounter is the pinnacle of wasteland reclamation. The 2.4 kilometre disused elevated freight railway track runs from the meat market district, parallel to the Hudson River, up to West 34th Street. The last train ran in 1980 and soon became rusted and overgrown. After twenty-three years of arguments between local residents, business and local government, a competition was set up to find a group of designers who could design an alternative usage for this monolithic structure soaring above the city. When James Corner Field Operations and Diller Scofidio and Renfro began work on the project, they faced difficulties of rotting wood, blood splatters and the smell of meat fat at the lift access points, old turnstiles, toxic soil and used condoms.[40]

The project reminds me of Goethe's *urform*, the pursuit of the perfect form, that basis or structure of all natural species ... except as a messy construct, a mimicry of the wasteland it replaced. *The Long line* is a brownfield[41] which is the term that refers to a site that has been previously used. Like Goethe's *urform*, the archetypal structure is there, it's just the details that have changed or have been variegated. The underlying structures of steel, wood, rivets, railings, beams and tracks were consistent along the entire railway line. Its form could not and would not be altered. It would be resuscitated as a slow passageway for an ambling walk. The designers were adamant there be no shops or cafes, nor areas to skate on wheels of any form. Feet only. The new development would change but without compromising its underlying form.

---

40   Corner, James, Scofido, Diller and Renfro *The High Line: Foreseen, Unforeseen*, London: Phaidon, 2015, 17.

41   Ibid, 302.

Wastelands are usually spaces that are potentially contaminated or are eroded. *The High Line* worked to retain the wasteland aesthetic by disallowing the kind of traffic that beach side or waterside boardwalks encourage. Instead, they designed for wildlife habitats and microclimates – wetlands, woodlands, herbaceous and wildflower fields. *The High Line* is a project where there are traces of the hubbub of its past life (trains swishing past at speed.) The hybrid hard surfaces were intended to slow traffic – the leisurely strolling of pedestrians wandering amongst artworks and flower beds would contrast against its past usage. *The High Line* is a revised or artificial wasteland and it is also a proto line that remains as a trace of its former usage.

The High Line is also a vertical 'haha.' A haha is a protecting line or zone between garden and agricultural or grazing land. In their 17th century French origins, they were originally sunken ditches with unscaleable walls on one side. In terms of its capacity to divide public leisure space and public utility space, the project is a haha. An open habitat that links two spaces, *the High Line* recalls this concept of the divide between wilderness and cultivation or civilisation.

In addition to being a haha between nature and culture, *The High Line* is a natural contract as per Serres' ideas of the natural contract where we are no more than renters on the planet. The natural contract presents a shift towards humans and all other elements of nature as having equal rights. The mosslands, woodland thickets, tall meadows and young woodlands of the project encouraged a consideration of the blurring of the line between plants and hardscapes. The designers wanted the plants to grow up through the cracks and over the edges of their designated areas. The designers said of garden designer Piet Oudolf's plants: 'The planting has such as sense of life – a buzz, a scent, a tactility – alongside decay ... The consistent systemic nature of the planking [hard walkways] is the perfect foil for the lyricism of the plants.'[42]

The High Line is not quite a garden, not quite its overgrown decaying predecessor. The length of greening blurs the borders between the disciplines of thought. Control or wilderness? Well, it falls between the two because keeping the plants alive without a never-ending earth beneath them leaves the plants vulnerable to freezing cold. But it is also an effort to leave the plants be, to allow them the capacity to be, but with an iron fist. *The High Line* is highly maintained and cared for, so it represents a paradox. It's a start, but it's not the end. In essence, it is a failed wasteland.

---

42   James Corner, Scofido, Diller and Renfro. *The High Line: Foreseen, Unforeseen*, London: Phaidon, 2015, 189.

The true wasteland appeals to us. It exists, in spite of its aesthetic displeasure, its disuse or lack of form. The attraction is so strong that *The High Line* worked tirelessly to mimic the qualities of the wasteland. The force of the wasteland also mirrors humanity's curiosity about existence outside cultural structures, outside moral codes. Humans like a bit of radical defiance. We like a place that won't be tamed. We appreciate the beauty in the ugly. Why? Because it suggests that a re-wilding (the plant contract) of a dark and vital plant realm, is still possible.

CHAPTER 2

Green Man: Human-Plant Hybrids

In Medieval times, nature held dominion over humans. Humans kept animals close, lived in modest dwellings and fished in rivers and bays, according to daylight and the seasons, weather and the tides. Nature was as much a source of worship as other, more religious, idols. Nature was closely affiliated to Medieval lore, folk stories and ways of life. Somehow, somewhere along history's tenuous line, the balance has become skewed and humans have lost the respect and fear they once had for the wrath of hail and flooding, for morning frosts that ruin crops and the cessation of work once the sun sets behind the far hills.

The Green Man is a Medieval image that sits strangely, but clearly, within a plant contract discourse because it is an overt reminder of what has been lost and of how powerful nature was once deemed by humans. The Green Man is an iconic carved stone (or occasionally wood) image of a face with foliage, many with leaves pouring from their mouths. As an artistic and architectural aesthetic, these stone carvings have adorned the bosses, eaves and buttresses of thousands of churches and cathedrals across the UK and Europe since the first century BCE and were particularly numerous during the medieval epoch. It was an image of pagan worship that was eventually commandeered by the Church, significantly in the 12th–15th centuries.[1] Green Man carvings, then, have existed for at least 2,000 years and can be seen, with hindsight, as a symbolic forerunner to recent philosophies of plant/human hybridity, where desire for closer connection with nature emerges from a state of loss.

The Green Man carvings share plant-human hybridity qualities with the work of several contemporary international visual artists who are also working with the vegetal. This chapter asks how culture-nature interactions in contemporary art now, as with historical Green Man motifs of the past, affect the desire to relate closely to plants. Any human yearning to connect with the plant world plays into the need for a plant contract. Green Man, in its hybrid mode, imaginatively collapses the distinction between the self and the other, and introduces the abject nature of human-plant hybridity.

The Green Man is the pin-up boy for a plant contract because it speaks of the war waged on the environment by humans. It speaks leaves. A history of the Green Man in Australia is limited to the last two hundred and thirty years of white settlement, with only a few extant examples – such as at the Adelaide

---

1   Richard Hayman. *The Green Man.* Sussex: Shire Publications, 2015, 5.

© KONINKLIJKE BRILL NV, LEIDEN, 2018 | DOI 10.1163/9789004360549_004

Jail, Hoskins Memorial Church in Lithgow and St Peters Cathedral in Adelaide, South Australia. Australian colonial artists struggled in their representation of the Australian bush, often returning to British conventions and imagery in their artworks instead. There was a lack of alignment between artworks created and the natural specifications of Australian bush vegetation: 'the fringe of the empire.'[2] In Australia, it is possible to argue that there has been an even greater yearning for plant-human hybridity as something that exists mostly on other continents, in other places far away. The focus here, however, is on documented Green Man carvings in the UK and on plant-human hybrid works by artists from the UK (Ackroyd/Harvey), Slovenia (Spela Petric) and the US (Edourado Kac).

Does the Green Man create a link between human and plant? How does it endure as an abject hybrid in the contemporary art works discussed? Does it remind us of nature – a mnemonic element that plays into the theory of a plant contract? First focussing on three artist case studies to interrogate the longevity of human-plant hybridity, the discussion is intended as a means of reclaiming human relations with the wilderness in this epoch of the Anthropocene. As Carolyn Dinshaw says, these medieval Green Man objects are important in contemporary life because they are helpful to 'think with' ideas of nature and they are deeply countercultural. Here, too, the Green Man helps make sense of our ontological relation with the vegetal world and functions as a reminder that desire for plant life (as source, as co-species) may be a reflection of how humans (especially in urban cities) have a reduced connectivity with extended areas of vegetal life.

Michel Serres, author of the natural contract, has more recently written of animism in his discussions of the relations between human and nature. To speak of signing a pact with nature, to assure Mother Nature that humans will wreak no more havoc, plainly demands an acceptance of animism.[3] If nature has become inanimate, isolated, removed from city life, distant from our public lives, then the plant contract must embrace animism in its attempts to return us to the source. The Green Man is nothing, if not animistic. Serres says, 'Love the bond that unites your plot of earth to the Earth.'[4] Green man is the bond.

The Green Man carvings were originally expressed as folkloric rascals; as apotropaic symbols to ward off evil; as pagan images of worship; and as signs

---

2   Bernard Smith. *Place, Taste and Tradition*. Sydney: Ure Smith, 1945, 3.

3   Serres, Michel. 'Faux et Signeux de Brume: Virginia Wolf's Lighthouse,' *SubStance* 37, 2, 116 (2008).

4   Michel Serres, *The Natural Contract*, Ann Arbor: University of Michigan Press, 1995, 50.

of bacchanalian revelry.[5] These stone features have been traced to as early as the 1st century BCE but the Green Man concept is also evident in German manuscript illustrations of the 12th century and in the work of such 17th century Italian artists as Pietro Ciafferi.[6] There was a cultural association of seasonal worship with the carving of these plant-human images. With each new season, fruit and berries would arrive, small crops would be harvested and herbs would be gathered. Nature in medieval times was considered something to appease, an entity of significant power and sovereignty.[7]

Green Man, its name changed from Foliate Face by Lady Raglan in 1939,[8] has also been cast by scholars as a figure of fertility, and therefore an evocation of bounty and fecund promise.[9] Were these wild men, the wodesmen, also used to explain unexpected pregnancies? In the small communities of country folk, the notion that a fertility god could bring such gifts from the forests was not so far-fetched. High up on church bosses, eaves and as interior adornment of the pillars and cornices, there are thousands of churches across Europe where the Green Man outnumbers images of Christ.[10] These high numbers of Green Man motifs ensured its presence as an image that was a mainstay of collective cultural knowledge across Europe and across time. Yet it is an image that eludes our easy understanding of them as mere fertility fetish or as offerings for ready harvests. Its hybridity, its morphing of two species provokes further thought.

Canon Albert Radcliffe of Manchester Cathedral points out there are thousands of churches across the UK with Green Man iconography. Norwich Cathedral has a number of leafy heads – nine visages with oak, maple, strawberry, buttercup or gilded hawthorn leaves. Richard Mabey reads one of these foliate faces as being a 'gigolo' and another being a 'diablo.'[11] He refers to them as 'symbolically sinful' and 'undoubtedly having a theological status.'[12] Mabey traces Kathleen Basford's research into 8th century theologian Rabanus Maurus' interpretation of Green Man's leaves as sins of the flesh from lustful and wicked men.[13] Mercia MacDermott focuses on the startling and grotesque Gothic iterations of Green Man, which have attracted the most scholarly

5   Tina Negus. 'A Photographic Study of Green Man and Green Beasts in Britain.' *Folklore* 114.2 (2003): 247–261, 247.

6   Brandon Centerwall. 'The Name of the Green Man.' *Folklore* 108 (1997): 28.

7   Richard Hayman. *The Green Man*. Sussex: Shire Publications, 2015.

8   Lady Julia Raglan. 'The Green Man in Church Architecture.' *Folklore* 50.1 (1939): 45.

9   Clive Hicks. *The Green Man: A Field Guide*. Fakenham: Compass Books, 2000, 3.

10   Ibid, 15.

11   Mabey, Richard. *The Cabaret of Plants*. London: Profile Books, 2015, 99.

12   Ibid, 101.

13   Ibid.

attention. William Anderson, however, sees the Green Man as a more generic universal figure of a Gaian connection to earth. Mabey refers to Kathleen Basford, a scholar who has worked with the history of the Green Man, as creating an 'admonitory' interpretation of the image.[14] In addition to being a scolding and a reprimand or a source of cheeky revelry, we can now see the Green Man, with hindsight, as a sad vindication (even an ironic validation) of our Anthropocentric disconnection from the natural world.

The Green Man gargoyles vary in shape, facial expression, leaf pattern and size. Leaves emanate from ears and noses in some. Others have leaves as hair and beards. Some have wolf-like foliage all across their faces. Others have clenched teeth or sad and soulful eyes. The most interesting, for me, are the Green Man carvings that have leaf foliage emanating from their mouths. They are spewing out leaves, 'expressing' back the nature that they are. They are speaking leaves: nature speaks through the Green Man. This is relevant in an era when plant science is showing us that plants sense and make decisions, learn and remember and yet we humans do not have the vocabulary to articulate what this might mean. Humans are limited by their anthropocentrism, necessarily, and are also limited by the incapacity to appropriately translate what plants are communicating. The Green Man reminds us that we have not allowed plants the agency and autonomy that is their nature. They remind us that humans are distinct from the vegetal world and that their expressions recall that there is no true hybridity, only an abject longing to hybridise. They are an example of how art can mediate our lost relations with the world (nature; plants).

The Green Man has been linked to the fourteenth century poem *Sir Gawayne and the Grene Knight*[15] which follows Arthur and his knights, who begin a mid-winter feast. An unknown knight arrives, dressed in grass green and riding a green horse with sparkling green stones along the saddle girth and an enamelled green harness. He carries a holly bough and a large axe with a blade of good green steel.[16] After an altercation, the Green Knight is beheaded by Gawayne; the knight then retrieves his head and rides away issuing a challenge to occur in 12 months' time at a 'grene chapelle.'[17] This literary reference supports the medieval curiosity in, and narrative effect, of the merging of man and vegetal greenness. The Green Knight is a stranger, a queer headless element, in an otherwise conventional tale. The Green Man

---

14   Ibid.
15   Hayman, Richard. *The Green Man*. Sussex: Shire Publications, 2015, 7.
16   *Sir Gawain and the Green Knight*. Trans. Jessie Weston. Cambridge: In Parentheses Publications, 1999, 7.
17   Ibid.

image later became associated with processions, pageants and revellers (men dressed in green and with foliage costumes) based on Joseph Strutt's engraving of the May Day figure, *Jack-in-the-green*.[18] There are connections we might tentatively make with 15th century outlaw Robin Hood and later figures such as 1962 Marvel comic hero 'The Hulk,' or even the silent root-man – the 1960 Marvel comic-book hero 'Groot,' who re-emerged in the 2014 and 2017 films *Guardians of the Galaxy*.

As an enduring hybrid icon, Green Man represents the human desire to merge with plants in order to safely moor humanity to the wilderness. Humans are a part of nature but also have a conscious cognition of it. The disconnect and connection between these two perceptions of the environment, as being immersed in it and aware of it, are evident in the motif of the Green Man. Human identity exists at the very point between nature and a constructed concept of it. Luce Irigaray and Michael Marder collaborated on a recent book, *Through Vegetal Being*, where they presented philosophical perspectives on vegetal matter and human being. They each responded, in individual chapters, to the same topics. Marder's philosophy seeks to raise the status of the plant, due to their generative capacities to seed and to sprout.[19] The Green Man sprouts a foliage beard and offers a foliate face and therefore might be considered a cultural figure worthy of greater vegetal consideration and ontological relevance – perhaps a champion of the vegetal world, that is receiving greater attention by Marder and other scholars.

In addition, the Green Man can be read psychologically as a way to fill this perceived lack or longing in humans. This longing, caused by a sense of exclusion from nature or *solastalgia*[20] is the distress felt as a result of changes to the green environment. Luce Irigaray calls this 'lack' an absence of thinking.[21] The lack, the exclusion and the loss also tie into Marder's 'plant thinking.' Plant thinking is not thinking like a plant or suggesting that humans have access to human-like plant thought. Instead, plant thinking is an ethical mode of making sense of the world, a 'vegetal deconstruction of metaphysics.'[22] Marder is referring to emerging plant ethics and plant behaviour in his field of Critical

18   Centerwall, Brandon. 'The Name of the Green Man.' *Folklore* 108 (1997): 28.

19   Irigaray, Luce and Michael Marder. *Through Vegetal Being*. New York: Columbia University Press, 2016, 135.

20   Albrecht, Glenn, et al. 'Solastalgia: The Distress Caused by Environmental Change.' *Australas Psychiatry* 15 Suppl. 1 (2007): S95–S98.

21   Irigaray, Luce and Michael Marder. *Through Vegetal Being*. New York: Columbia University Press, 2016, 3.

22   Ibid, 113.

Plant Studies, and how this changes the way humans think, ontologically and epistemologically, about being human and knowing the world as humans. Marder's current influential work and the enduring relevance of cultural icons such as the Green Man begs the question: at what point did humans, mostly in the Western urban world, become separated from nature, yet wish so vehemently to reconnect? And while we might rush to bookend the Age of Reason, and the Romantic period, the Green Man is important as a reminder of changes in human perceptions of nature. Where the foliate face once was associated with a respect for, and worship of, the natural processes of the wilds, it now is a reminder of the widening gap between the natural world and human perception of it. To desire nature, to be in it, with it, amongst it, to eat it, be protected by it and to represent it is a desire, charged and sexuate.[23] This desire can be seen as an acknowledgement of the differences between human elements and distributed plant behaviours.

The first artist discussed is distinctly aware of the vegetal qualities that might be shared between human and plant species. In the Green Man we see extrusions of foliage from mouths, part of the cultural aesthetic of a time, which was deeply connected to the prospect of a merging of human and plant life. US artist Eduardo Kac works with plant-human material, which is of interest in new Critical Plant Studies. Kac creates the abject in a distinctly hybridised plant-human way by casting off singular human qualities and experimenting with a merging of the human with the plant. Kac genetically engineered a flower that was a hybrid of a petunia and himself. Developed between 2003 and 2008, and exhibited at the Weisman Art Museum in Minneapolis, this abject artwork – *Natural History of Enigma* – created a new life form called 'Edunia.' Kac called his flower a plantimal and he described the genetically engineered flower as expressing his DNA exclusively in its red veins.[24]

Kac describes his work as follows:

> The new flower is a Petunia strain that I invented and produced through molecular biology. It is not found in nature. The Edunia has red veins on light pink petals and a gene of mine is expressed on every cell of its red veins, i.e., my gene produces a protein in the veins only. The gene was isolated and sequenced from my blood.[25]

---

23   Irigaray, Luce and Michael Marder. *Through Vegetal Being*. New York: Columbia University Press, 2016, 87.

24   Edourdao Kac. *Signs of Life: Bio Art and Beyond*. Ed. Eduardo Kac. New York: MIT Press, 2006, 42.

25   Ibid, 43.

Kac's Edunia can be propagated through its seeds but each time a new flower sequence occurs, his own DNA will be present. Whilst medieval cultures might have seen the Green Man as a reminder of potential ills, or a facilitator of fertility, the Green Man in this example reflects a human need to merge with nature. A dangerous need, it is a wilful desire that Kac has likewise exacted with the Edunia.

Desire as Longing

Mabey suggests, after all the Green Man gargoyles he has seen, that these foliate faces were 'caricatures of village elders, terrifying portents of damnation, clever visual puns ... [and] that over the centuries they developed into an all-purpose design feature, a logo endowed with the perennial magnetism of the chimera and an irresistible eye-worm for stone carvers.'[26] Kac's work has a solemnness and a sobriety that does little to recall the lascivious Green Man gargoyles. Like the Green Man icon, however, Kac's scientific merging of two species is an expression of desire and an abject need to remove the barriers between human and plant species. A desire for nature can act as a call to life. As Irigaray says, 'Desire, then, loses its living roots, unless it falls back into more or less wild instincts.'[27] In other words, lack of contact with the natural world creates a deficit: a lack of life, a lack of living. Better, in fact, to become re-connected with sexuate beings in order to connect better with the other-than-human wild world: 'such a cultivation of our sexuate surges is crucial for our becoming able to behave as a living being among other different living beings without domination or subjection.'[28]

The head of the abject Green Man might fit well into Julia Kristeva's concepts of ejecting death. Rather than speaking a non-human tongue, perhaps the Green Man is rejecting mortality by spewing forth or rejecting that end. As Kristeva says, the abject is neither subject nor object and the abjection is the casting off of the self as well as a perverse rejection of the sacred.[29] This is relevant in the sense that it is speculated that the Church

26   Mabey, Richard. *The Cabaret of Plants*. London: Profile Books, 2015, 102.
27   Irigaray, Luce and Michael Marder. *Through Vegetal Being*. New York: Columbia University Press, 2016, 86.
28   Ibid, 87.
29   Kristeva, Julia. *Powers of Horror: An Essay on Abjection*. New York: Columbia University Press, 1982, 26.

adopted the image of Green Man to attract the pagans to its pews.[30] Yet there is an anti-religiosity to these images. To be afraid of the Green Man is to be afraid of the unnamable; fear is 'the frailty of the subject's signifying system.'[31] Kac's work, too, is frightening. This is not the hoped-for connection to the wilderness. Instead, Kac's hybridity is cultivated, cultured in a laboratory, combined in a petri dish. This is a rewilding of culture but it is a queering of that desire to re-wild.

The Rewildings

A plant contract is a form of rewilding. The natural world cannot be human property. And yet humans 'own' most of the land upon the earth. At least, humans have paid vast sums to own a deed or land-sale contract that tells them they hold ownership of a pocket of earth and air. The rights and ownership of the land is a mastery that cannot hold firm. Vegetal forces are slipping in between the established patterns of being. 'The Earth existed without our unimaginable ancestors, could well exist today without us, will exist tomorrow or later still, without any of our possible descendants, whereas we cannot exist without it. Thus we must indeed place things in the center and us at the periphery, or better still, things all around and us within them like parasites.'[32] There are artists who are shifting the focus from human parasite to Earth. They are playing with the way nature grows, how species rely on one another in symbiosis rather than parasitism and the way our conventional understanding of the natural world can be radicalised.

Another contemporary example of hybridity and abjection involves the dismantling of boundaries between human portraiture and grasslands. Like Kac, Heather Ackroyd and Dan Harvey are using new cultivated species of plant and artificial modes of horticulture to metaphorically re-wild human lives. Ackroyd and Harvey are British artists, each of whom were working with grass when they first began collaborating in 1990. They undertook study of particular grass species in Wales. Ackroyd and Harvey have made a sensational art career from working with grass. They filled churches, stone staircases, gallery walls – and erected five-metre-high curtains – all using grass. They use the process of projecting negative images of a human face on walls of grass they have grown, using the light to manipulate the natural process of

30   Hayman, Richard. *The Green Man*. Sussex: Shire Publications, 2015, 5.

31   Kristeva, Julia. *Powers of Horror: An Essay on Abjection*. New York: Columbia University Press, 1982, 34.

32   Michel Serres, *The Natural Contract*, Ann Arbor: University of Michigan Press, 1995, 33.

FIGURE 4   *Dan Harvey and Heather Ackroyd. Miles, Bashia, Nath and Alesha 2007.*
             GRASS, PHOTOGRAPH.

photosynthesis (see figure 4). They have said, 'Grass may be the material of our investigation but chlorophyll is the primary medium that binds us.'[33]

The exciting albeit abject aspect of the duo's works is the way the image emerges and 'grows,' merging the human face with the growing stems of grass as an artistic expression of a casting off of the self. At last we see the human temporality of growth at the same rate as the vegetal temporality of growth. From 1997, Ackroyd and Harvey have been collaborating with scientists Helen Ougham and Howard Thomas, who work in pioneering new strains of grass. The interesting and yet so obvious element of their work is that they grow grass from seed on the vertical plane. This makes shadowing and the development of photographic imaging more effective. The various shades of yellow and green light work consistently as a black and white photo might. By projecting a negative photo image onto the seeds as they grow and manipulating the light in that way, the image forms:

> [O]nce exposed to light in a gallery environment, the grass in the yellow
> regions quickly seizes the available light and gradually, over hours,
> changes color, greening up. Kept in very low light levels in a living state,

---

33   Ackroyd, Heather and Dan Harvey. 'Chlorophyll Apparitions.' *Signs of Life: Bio Art and Beyond.* Ed. Eduardo Kac. New York: MIT Press, 2006, 3.

the green grass begins to dismantle its chlorophyll and, taking on a quality akin to an old tapestry, the image slowly fades away.[34]

The 'lack' displayed here is less an emptiness or a futile hopelessness about the current environmental concerns and more about the drive to fill the lack. These are the desiring motivations that urge us to return to nature or merge into a hybrid state, to create lines of desire that deviate from the civic passages through our urban and non-urban landscapes. As Irigaray writes, 'Desire can transform a simple territory into a world.'[35]

The last plant-human hybrid artist to discuss is Slovenian Spela Petric who 'confronts vegetal otherness' in her performance work *Skotopoesis 2015* (also performed at the *Click Festival* in Denmark in May 2017). Petric created a nineteen-hour performance work that involved her interaction with germinating cress, a more formal bodily hybridisation. Her experimentation was with ethics and inter-cognition between plants and humans, and the work explored the 'novel human-plant relationships beyond the limits of empathy, interfaces and language.'[36] Her work is relevant to the idea that the Green Man concept of hybridity and abjection has endured into contemporary aesthetics. Like the Green Man, there is evidence of a collapse of boundaries between self and other, between human and other-than-human environment, between artist and cress plants. This collapse creates a disruption, ideally, so that viewers start to see human relationships with the plant world differently because we become more aware of the influence of one species upon another, rather than relying always on the effects of human life on the vegetal world alone.

Skotopoesis 2015 (see figure 5) was a durational work where Petric stood with her shadow falling across a tiny field of green cress plants. She wanted to achieve a Barad-like agential cross-over between species (American feminist Karen Barad has led the way in the theory of agential realism). Petric used 400,000 cress seedling plants. She notes that the cress plants were aware of her form as a presence that caused a shadow and prevented their access to light and therefore photosynthesis. Petric wanted to achieve 'an economy of mutual suffering.'[37] She was uncomfortable having to stand there for such a

---

34   Ackroyd, Heather and Dan Harvey. 'Chlorophyll Apparitions.' *Signs of Life: Bio Art and Beyond*. Ed. Eduardo Kac. New York: MIT Press, 2006, 2.

35   Irigaray, Luce and Michael Marder. *Through Vegetal Being*. New York: Columbia University Press, 2016, 95.

36   Petric, Spela. 'Confronting Vegetal Otherness.' *Skotopoiesis*, 1. 6 December 2016 http://www.spelapetric.org/portfolio/skotopoiesis/.

37   Ibid.

FIGURE 5   *Spela Petric. Skotopoiesis 2015. Artist and germinating cress.*
PHOTOGRAPHER: MIHA TURŠIČ. COPYRIGHT: ŠPELA PETRIČ.

long time as the areas of the cress that her shadow fell upon suffered lack of light. Of course, the ethics of this is spurious because the plant had no 'say' in the matter. The blades of the cress might lean towards the edges of light on the outskirts of Petric's shadow but ultimately the artist exerted an influence over the situation. This was the breach of ethics that the performance was intended to provoke.

Petric says 'vegetal ontology is alien to humans which prevents us from establishing a legitimate empathic relationship with them.'[38] She suggests it might not be human ignorance that allows us to think it is legitimate to use non-human species as manipulated subjects and usable. There might be a more fundamental metaphysical problem. Her reading of Marder's 'Vegetal Metaphysics' paper is commensurate with his conclusion that empathy for plants disregards their mode of being.

Skotopoesis, for Petric, means *as shaped by darkness*. She notes that as her water cress grew as the result of the artificial light, she began to shrink, from dehydration compressing her vertebrae. The audience was able to observe and, afterwards, ask questions. Once the performance was over, Petric asked that the cress be harvested and eaten by the gallery staff. Unfortunately, this

---

38   Petric, Spela. 'Confronting Vegetal Otherness.' *Skotopoiesis,* 1. 6 December 2016 http://www.spelapetric.org/portfolio/skotopoiesis/.

was not done, for reasons never given but most likely relating to the rush of having to de-install exhibitions under tight timelines to make space for the next exhibition. Irrespective of why the cress was not dutifully eaten, it ended up being thrown away. This is ironic in the context of an experiment that was intended to unmask the ethics of how we engage with plant life. Perhaps it is time to redress how we relate and interact with the vegetal world and Petric has, at the very least, contributed to that debate.

The connection between the Green Man and the burgeoning interest in plant matter in contemporary art has yet to be fully resolved. The reasoning, here, is that it is related fundamentally to desire and the abject. Art might be a representation, a copy, a memory or an expression of nature; it might be a response to natural beauty or to nature's creativity and production. But art is always a *reminder* of nature.

In the Green Man we see this non-distinction between the self and the natural world. Green Man represents the merging between species and because Green Man carvings still exist in churches across the UK, time allows us to attempt to understand their relevance when compared to contemporary art works in the epoch of the Anthropocene, as well as in the context of current Critical Plant Studies. This conflation of both time and plant/human expression creates the context in which these artworks have been discussed. What Green Man and the three contemporary artworks have in common is a hybridisation, yes, but they also share a common quality which is that something is not quite right. The outputs are, frankly, a little unusual and decidedly queer. As Dinshaw says,

> These aesthetically intricate, affectively intense images represent creatures that are strange admixtures, weird amalgams: they picture intimate inter-relations between the human and the non-human – interdependencies between species that throw taxonomies into question, rub categories up against one another, put classifications and hierarchies of the human under scrutiny. These are queer creatures indeed.[39]

Dinshaw is interested in the relation of human to non-human, tracing the afterlife of this medieval image to imagine more expansively queer worlds. To allow for sexuate differences and ambiguous relations, to indulge our desires for closer connections to the plant world (or other non-human worlds), we must be prepared to lose our sense of sovereignty over nature. The Green Man

---

39   Dinshaw, Carolyn. 'Black Skin, Green Masks: Medieval Foliate Heads, Racial Trauma, and Queer World-making.' *The Middle Ages in the Modern World.* Ed. Bettina Bildhauer and Chris Jones. Oxford: Oxford University Press, 2017, 1.

emerged in great numbers in medieval Europe and endures today, unlike the forests and bushlands from which they supposedly emerged. This lingering image, then, serves as an important reminder of loss and lack. By this, I mean they remind us of how the environment has been devastated at human hands and that our desire to connect with nature is as strong as ever. The gap between the two is where queer and abject art is being made. The plant contract may have these queer and abject attributes too, for the same reasons.

The Plant Radicants

With Green Man or Foliate Face, it is possible to see how humans have been consumed by ideas of plant-human hybridity. The use of plants for medicine, narcotics, food, oxygen and building materials fuels the dynamic relationship between human and nature. We rely on nature, grow it, harvest it and worship it. Despite our more recent global capacity to destroy nature, humans have a fundamental yearning to be closer to nature, to hide in it, tramp in it and to seek succour amongst it. This paradox makes the study of Green Man, in the context of contemporary art being made within Critical Plant Studies, the desire to be one with the vegetal world, a pressing issue in our epoch of climate change.

The question discussed in this chapter is how art acts as a mediator between human and plant in this discourse of hybridity and whether these connections are useful as a means of disseminating ideas about human relations with the world. A philosopher with a track record in art criticism is Nicolas Bourriaud. He suggests the radicant is the vegetative figure, an artist working within a context but destabilising that context at the same time. The artists chiseling the green man in medieval architecture's gargoyles could, in this context, be seen as radicants. So too could Kac, with his experiments with petunias, or Petric with her cress shadowing. Bourriaud emphasises that the artist in the 21st century must reject the 'standardisation of imaginations decreed by economic global-ization. To be radicant: it means setting one's roots in motion, staging them in heterogenous contexts and formats, denying them any value as origins ...'[40] What postmodernism calls hybridisation involves grafting onto the trunk of popular culture that which has become uniform markers of 'specificities' – features, usually caricatured, of a distinctive ethnic, national, or other cultural identity – just as mass produced candies are infused with different synthetic

---

40   web site on Sternberg book's page: http://www.sternberg-press.com/index.php?pageId=1
      224&bookId=119&l=en.

flavours.[41] Bourriaud's concepts of transplanting is the constant discourse between artworks and their contexts. This has a relevance to this text's efforts to draw connections between artists who reclaim plant life to mediate issues of climate change. Where Bourriaud advocates exchange of ideas, this chapter charts the materiality of an exchange between plant and human. Vis-à-vis: the Green Man.

If nature cannot be separated from culture or from the human, then hybrid monsters are a logical source of interest. This chapter sought to investigate the plant-human hybrid, in the context of a precarious natural future. The contemporary artists already discussed are, in Bourriaud's terms, performing acts of translation.[42] Art can translate difficult issues in society, and the lowly status of plants and vegetal thinking is exactly the kind of issue that is difficult to generate in mainstream discourse. For this reason, the impact of art on the dissemination of new plant science is critical. The power of plants is their reminder of our foolish ways, our hubris, our relentless desire for more. How can we be so careless about our mortal lives?

Deathly Monsters

An aspect of the original Green Man motifs that has not yet been discussed is the issue of death. The original stone motifs are a merging of plant and human. I think it is important here to suggest that this merging of the human with vegetal leaves might be a *return to earth*. It may be a dust-to-dust acknowledgement of the burial of the dead in the earth. There – where the soil turns flesh to dirt, where the leaves and plants grow over the grave – that is where all humanity ends. The composting of human and plant into one organic matter.

The horror of death and our human inability to outrun it, then, could be a second significant part of the Green Man discourse. Rather than Kristeva's ejecting of death, as discussed earlier in the chapter, this represents a reading of the Green Man as an openness to death, a welcoming of the processes of life and death, decay and deterioration. But death is never the end. Finitude outlives the kinds of deathly experiences we can ever experience or witness in our 70–80 years of life. Apprehending death is to consider the monstrous in us.

---

41   Nicolas Bourriaud, *The Radicant* New York: Lukas and Sternberg, 2009, 20. http://www
.mammalian.ca/pdf/Radicant.pdf.

42   Ibid.

There is a tradition of plants and humans becoming a ghastly hybrid at the nexus of science and the humanities. Monstrous images of human-plant interactions are mostly evocations of the strange 'other.' This may be a response to the unusual species that already exist in nature, which appall us with their fearfulness and fluid sexuality, their rampant strength and their unfamiliar qualities. Nature is brimming with examples. For instance, in a recent newspaper article, a single honey fungus mushroom in the Oregon Blue Mountains occupies nearly 2,000 football fields. Each individual specimen is as massive as this. There has been much research on the benevolent and complex communication information systems of subterranean mycelium, yet these more malevolent honey fungi are parasitic, the fruits of the rhizomorphs and hyphae below ground. They destroy all trees and plants in their paths.[43]

The monstrous plant creature, alive and wild, is not to be relegated to the real world alone but is represented in creative ways in literature and film. By creating an unruly creature, we perpetuate the Romantic concept of nature as being something that exists as separate from the human. But in these instances, the wild thing is a merging of species and bridges the philosophical gulf between subject and object. It becomes something that is neither subject nor object – it is abject.[44]

The Swamp Thing was just such a thing. It was a character in a story that had many incarnations. The 1987 saga was my favourite version.[45] It was a comic book story that followed a terrible laboratory experiment, where Dr Alec Holland turned into a plant-human hybrid after an explosion in his lab. This character was a reinterpretation of a former minor comic character, Floronic Man. Holland became a humanoid mess of vegetal matter. Swamp Thing's face fills up with rain as he changes consciousness from human to plant – so ghastly. Despite his new monstrous form, he strives to protect his swamp home and to preserve the community from various threats. There is genuine ecological concern for the world in this comic book hero.

What is this wish, by the writers Moore, Bissete and Totleben, to recognize the plant in our own faces? Can we see, in the vegetal visage of a plant man, the truth about the way we are? Alphonso Lingis says, 'Coral fish, butterflies and wasps, birds of paradise and hummingbirds, zebras and foxes bear surface colours and patterns and utter distinctive cries with which they both recognize

---

43   Elzbieta Sekowska, 'Meet the world's Largest Living Organism,' *IFL Science*. http://www
      .iflscience.com/plants-and-animals/meet-worlds-largest-living-organism/.
44   A neologism, despite these being very unpopular in all disciplines.
45   A. Moore, S. Bissete and J. Totleben, *Saga of the Swamp Thing*, Vertigo, 1987.

one another and are drawn to one another.'[46] We are charmed and captivated by a face that is known to us. But the Green Man is wild, he lives in the forest and must be placated with offerings and prayer. The Swamp Thing was a clever man who dabbled with the dark arts of the science lab one time too many.

The Green Aesthetic

There are a number of other artists working in this interacting area of plants and human desire, at the point of biology and its microscopy. Similar to the scientific elements of Kac's work, there is a science laboratory aspect to hybridity that is darkly attractive and which has been taken on by others. The biological merging of man and petunia, the explosion in Dr Alec Holland's lab: these are the mad scientists, working in closely controlled facilities with the latest technologies.

Lea Kannar is an artist who was drawn to the petri dish process of working with plants. She works with the tree daisy or Scalesia pedunculata as part of her art practice.[47] She has undertaken research in the Galápagos Islands, finding out details of this community plant species. According to Kannar, there are groves where the trees are the same age, and which can thrive or fail as a grove. She encountered several varieties of daisy bushes and shrubs on the Galápagos Islands, however the tree daisy can only be found in the highlands of Santa Cruz in the Galápagos. Kannar's interest in trees also extends to the tree dandelion, a native of the Canary Islands. The dandelion is often thought of as a weed, growing up on verges and inhabiting neglected spaces, but is also used for its beneficial functions in gardening. It is also used for tea and coffee, and other topical medicines.

Kannar was able to fulfil a residency at the School of Visual Art Lab in New York in 2013 and 2014.[48] She says, 'It is a lab designed for artists to work with the assistance of qualified botanists (Sebastian Cocioba), marine biologists (Joseph DeGiorgis) and other specialists such as Dr Brandon Ballengee (artist and frog and marine specialist who was working there during her two residencies).'[49] This unique situation gave Kannar the capacity to explore areas of art that are usually difficult to find, where she was assisted to create her artwork and not

---

46   Alphonso Lingis, *The First Person Singular*, Northwestern University Press, 2007, 67.

47   Prudence Gibson, interview with artist Lea Kannar, email, 23 December 2015.

48   http://bioart.sva.edu/.

49   Prudence Gibson, interview with artist Lea Kannar, email, 23 December 2015.

trying to get the work done whilst staying out of the way of the scientist in their lab.

Suzanne Anker, director of the Visual Art Lab, and her assistant Tarah Rhoda regularly work to make sure the artworks can be of the highest standard. During the residency they also have talks and workshops with places like Genspace and visiting artists like Professor Victoria Vesna, Ph.D. professor at the UCLA Department of Design Media Arts and director of the Art|Sci Center, UCLA. The VCA art lab is best known for helping bio-artists such as Kannar who is interested in microcosmic cell work (in this case the cells of the Sonchus canariensis, Galápagos Tree Dandelion) to create artworks. She uses ceramics, prints, photography to express this cell imagery.

Like the previous artists mentioned, Kannar is interested in the minutiae of plant life – growth, replication, movement and sentience. Her work is an example of how the real life of plants can be expressed to a wider audience via aesthetic practice.

Conclusion

The Green Man is always male. There are no examples of a Green Woman. The female version of Green Man has been suggested to be Medusa.[50] The daughter of Gaia (earth) and Oceanus (ocean), Medusa and her hair of venomous snakes makes an interesting sister character for Green Man. Both are spawned from the earth. Medusa's fearful characteristic, however, were borne from her true and undying love for Poseidon. A tragedy of sorts. Perhaps a more likely sister image is Sheela-na-gig a similar entity (an image of biology/female sexuality) who was also an architectural apotropaic symbol to ward off evil. The challenge was to put your hand inside Sheela-na-gig's hole! Placing your hand inside a darkly female crevice within a frightening visage is not so easy.[51] Medusa and Sheela-na-gig, unlike the Green Man, do not seem to be speaking. There are no emissions from their mouths, however there are examples of the Green Man where leaves emanate from its mouth. It utters branches of leaves rather than words. This is no staccato. This is no intermittent conversation but a never-ending monologue. Or perhaps the Green Man is uttering a prayer, incanting a hymn of praise to its two creators, the artisan that modeled the stone and the wilderness that he is now no more than a memory of. We can only speculate.

---

50   https://www.greekmythology.com/Myths/Creatures/Medusa/medusa.html.
51   For Sheela-na-gig images, go to http://www.sheelanagig.org/wordpress/.

The endless twirl of leafy branches exuding from Green Man's mouth is a string of utterances. A series of leaves and stems and buds, of lobes and sinuses, of petioles and axils. What looks to the human eye to be a bit of greenery is in fact a constant flow of complex elements. These flows of foliage are like strings of linguistic objects. They are signifier sounds of information. It is an arbitrary language, a strange plant communication outside our general cognition. The stonemasons were in control of the force of this kind of imagery.

The Green Man: it is both a representation of fearfulness about the woods and the wilderness but also a desire to become one, to merge, to reconnect and to linger amongst the dark trees where nothing is certain. The artists I have mentioned in this chapter such as Spela Petric, Ackroyd and Harvey, Edourdo Kac and Lea Kannar are experimenting with that hybridisation of plant and human. This chapter has drawn on the mystery of the Green Man and chased him into the hearts of contemporary artists who are willing to share their DNA, impose their shadows, print photographs on grass and work together with plants to change our perception of ourselves … as hybrid human-plants. In these works, there is a longing to connect with vegetal elements, a desire to connect with a universal form – plant life.

Whilst the images may be classified as men, there is a chance to reclaim the concept of the plant-human that defies gender specificity. More important than a duality of sexuality, the Green Man is reminder of the human habit of trying to fill the lack, to replace the void. The Green Man reminds the church-goer of the unknown, the fearful and the abject. It reminds us of the fluidity of being, the closeness between species rather than discrete separations. This flow and leaking between species is at the heart of Green Man's being.

Monstrous, often ugly, the Green Man can now be seen as a casting off of agnotology, that habit of purposeful or willful ignorance. The Green Man can be reclaimed as a reminder and as a warning, not of the dangers lurking in the forests but the dangers of not taking care of the very same forests. A melancholy Serres says, 'We no longer know the world because we have conquered it.'[52]

---

52   Michel Serres, *The Natural Contract*, Ann Arbor: University of Michigan Press, 1995, 35.

CHAPTER 3

Robotany and Aesthetics

The humanoid robot poses many questions regarding 'cross-species becoming' and artificial intelligence. Humanoids also highlight changes in our relationship with nature and technology, and our human-computer interaction with elements of both (and perceptions of both). I've met one of these human-like droids.[1] In a robotics lab of an art school nearby, she fluttered her eyelashes and mumbled *rhubarb, rhubarb* at me whilst cocking her head from side to side. Cheeky devil. It was both exciting and disappointing, possibly because our expectations of robotics are still too high – overly informed by films and sci fi books rather than reality, as we are. Humanoids are technological provocations about who humans are and how we live in the world and how we see ourselves. Robotany – the connections between robots and vegetal life – in many ways, poses similar questions and extends the robotic into the vegetal realm. Vegetal life, up until the Enlightenment, has been cast as mechanical and without consciousness, but that identity of mere instrumentality is changing. So are the distinctions between the two. Robotany probes issues of human desire for plant life, to merge, to become one, to seek the original form.

Contrarily, and for so long, we have considered both plants and technology as mere utilities and this text seeks to redress this error, because we are living in the third media revolution where technology and nature and the human are converging. Plants and technology are performing subjectivities not inert or non-agented objects.[2] We are dying from the self-sabotaging process of removing ourselves from nature.[3] It seems that we will continue this slow death if we persevere with an endless disconnection between nature from culture. Robotany is a solution to this deathly separation as it comprises the ultimate hybrid being. These kinds of innovative solutions are intrinsic to a plant contract, which not only reminds us of the human war on nature but disrupts the progress of these violent atrocities.

---

1   In 2012 at the University of New South Wales Art and Design Creative Robotics Lab, Geminoid F, created by Professor Hiroshi Iriguro, from Osaka University.

2   Murphie, Andrew. 'Convolving Signal: Thinking the Performance of Computational Process-es,' *Performance Paradigm*, 9, 2013.

3   Serres, Michel. 'Faux et Signeux de Brume: Virginia Wolf's Lighthouse,' *SubStance* 37, 2, 116 (2008): 129.

Plants are a source of food, of shade, ready to be cultivated and harvested, grown for human consumption and aesthetic enjoyment. Now that we are discovering that we need a more appropriate philosophical approach to plant (and all-species) ontology via a plant contract, and now that these ideas are supported by scientific evidence about vegetal capacities, we are in a better position to change our attitude towards, and perception of, plants. When you also throw technology and robotics into the plant ontology mix, the game changes once more. Another click of the compass.

Plants are not static or fixed. Plants are not passive automatons.[4] They are constantly moving, growing, morphing into different iterations of themselves. Each part of a given vegetal element is not made fully and then left to function as part of an overall body. Instead, plant parts are each evolving and changing, as dynamic beings. Much of the technology and/or robotics that is discussed in this chapter refers to the amplification and augmentation of plant life, with an eye towards aesthetics. But it is important to remember that these improvements are human-focused: anthropocentric. Humans are using technology to support plants for humans – for human causes and for human production. This might at first appear to contravene a plant contract, as it seems to be a partisan position. However, I am arguing that technology and the human and nature itself are parts of the whole. Likewise, technology and the human are immersed in and created from natural elements. In other words, whilst it is human nature to create discrete distinctions between parts, this book also seeks to reassemble the elements into a Gaian-like whole.

Aesthetic Problematics

Robotany ties into the plant contract as part of a contractual collectivity. According to Michel Serres, a nature contract must extend to different disciplines such as industry, government, community etc. Robotany develops that collectivity to the realm of media and technology. Robotany, in this book, refers to works of art that incorporate robotics and plants, kinetic technology and plants, robotic augmentation and plants, and sonic synthesis of plants – within an art aesthetic.[5] Lucy Suchman refers to robot-child-primates as

---

4   Matthew Hall 'Plant Autonomy and Human-Plant Ethics' *Environmental Ethics* 2009, 9.

5   Other robotic or sonic plant works, not otherwise mentioned in the body of text of this chapter are: *"Plantas Parlantes"* (2010) by Ricardo O'Nascimento, Gilberto Esparza, Javier Busturia, Jigni Wang http://blog.derhess.de/2014/06/14/biosensing-for-human-computer-interactions/; Ben Kolaitis – biosensing at Serial Space."*Noisy Cauliflower*," 2013, by Cara Stewart; FoAM – robotics work; Ivan Henriques – "Jurema Action Plant"; Chiara Esposito – plant robot.

'almost human'[6] as a way to grasp the changing capacities of natureculture hybrids. In a similar way, I draw a conclusion regarding the art-plant-robot – that is, that they sit at the 'almost human' end of the 'becoming human' spectrum. If we can avoid seeing the robot as substitute for the human, then the roboticisation of the plant, within art, can be seen as a process of mutual activation and inter-dependence between species, and not substitutions for nature.

Where Suchman imagines possibilities of robot-merging, I am interested in the aesthetic issues and paradoxes that emerge from this kind of media art work, such as non-aesthetics – an inability to perform politically or move beyond mere representation, which relies too heavily on recording and mimicry, without aesthetic rigour. Where Suchman notes forms of natureculture to reveal how species are entangled, I am interested in how we have not escaped our aesthetic habits of imposing our humanness upon nature, even with future technologies at hand.

Aesthetics in this environmental and technological domain creates difficult terrain to travel. The aesthetic criteria for success of art that incorporates plants and plant/nature philosophy comes under scrutiny in this arena, because technology sometimes obscures or makes invisible the rigours of aesthetic process. I propose that it is timely to devise new forms and methods of evaluating effective works of art that include plant life, art and technology. Many artists working in this area are so consumed by the mechanics and methodology of their pseudo-scientific inquiry that they have left aesthetic criteria languishing back at the starting point. It is possible to argue that aesthetic success has not consistently been the focus of much of this genre work. What might the appropriate criteria be? An empathic experience of the vegetal? A creation of cross-over points amongst species? An intense artistic simulation of the life of a plant? A new awareness of critical plant studies issues? Perhaps all of the above.

Humans have a long history of exacting the processes and criteria of aesthetic judgement.[7] Pre-20th century approaches to aesthetic value affirmed that

---

6   C. Casteneda. "Robotic Visions" *Social Studies of Science* 44,3, (2013): 1–27.

7   Plato, in his *Symposium*, advised humans to be just, courageous, temperate and to have wisdom before they make aesthetic judgements. Kant's assertion was that judgement of the art work must not be preceded by expressions of the pleasure of the artwork. Kantian judgement must be rational and intellectual and immediate pleasures, while valid, are a distraction from the formal processes of reasonable judgement. Pleasure or the 'mere pleasantness of the sensation' only allows for private validity and does not inform his notion

aesthetic effect must be rational and reasonable as well as sensory. Modernist notions of the artist as genius, and then postmodern concepts of art as an aggregated entity (ironic, a pastiche, a fragment) helped to develop a different thesis for the aesthetic. Plant art in the 21st century requires a novel spectrum of criteria. It does not fit into any of these previous aesthetic theories. Plant art must fall within a category of ecological art, except there are no clear standards for such a category. This need for a completely new system of thought was experienced nearly two hundred years ago by the German idealist thinkers such as Goethe, von Humboldt, von Neumeyer and the various other botanical explorers of the Enlightenment. They were so concerned with taxonomy, category, classification, archetypal form and the rules of aesthetic representation, and they were so governed by mimesis, that it became difficult for many colonised countries to establish painting practices appropriate to their own environment. Australian colonial era painting is an example of this inability to move beyond classical art training and a subsequent obstacle to truly responding to the bizarre and unwieldy plants of the Australian bush.

We face a similar issue now. New discoveries in plant science throw us into the same kind of turmoil – both philosophically and aesthetically. How should we express the natural world, now that plants and trees display subjectivities rather than the mechanical behaviour of non-sentient objects? What do we do with these burgeoning areas of Critical Plant Studies and how do aesthetic critique and judgement fit into this plant contract discourses?

The Robotic as Criteria

If you have ever spent time watching the leaves of a fig tree flicker in a breeze, then you have seen that the branches and leaves respond in similar ways each time the wind picks up. You could almost say the responses of the leaves to the environmental stimulus are *robotic*. When we use this language, we are assuming that robots have a set of stimulated, repetitive responses, programmed to react within a small array of automated operations. Most fig trees have a limited and repetitive manner of growing, moving and decaying,

---

of universal attractiveness or sensus communis. Kant also felt that pure judgement must be independent of charm and emotion. He defines taste as contemplative pleasure. This is different from his practical pleasure which is bound up with desire (either cause or effect). Kant warns against 'all interest' because 'it spoils the judgement of taste.' (p. 72). Edmund Burke reminds us that aesthetic principles of art are embodied entanglements and that the spectator experiences feelings of love, when the work is very good.

yet the differences between nature and technology are still at extreme ends of the aesthetic construct, kept discrete by the touch of the human hand. This chapter, however, looks at artworks where nature and technology are not mutually exclusive.

A limited view of robotics endures (as it has with nature until recently). A generous dose of paranoia about AI matches the reductive thinking about robotics, which creates a paradoxical force – this will be made clear in several case studies of artworks discussed in this chapter. Suchman's curiosity in the imaginary possibilities of robots is relevant here, as it acknowledges the constantly changing relations between humans, robots and other biotic species. In the context of recent critical plant studies, the idea of plants having non-automatic capacities and imaginary capabilities is important, especially at the vector of art creation. Perhaps then the 'robotic' in 'robotany art,' referring to the technological augmentation for mechanised processes, might be an important aesthetic criterion. Does the artwork utilise technology to enhance its efficacy, impact and experience? If yes, then it can be added as a value element in the process of aesthetic judgement. The same can be said for the artwork's concept, reach and ability to change the viewer's attitudes to the given issues it arouses.

The recent artistic interest in plants might be seen as a running away from extinction, a fleeing from the real demands of environmental crisis. Moving towards plants for answers might be no more than an easy theoretical 'nature escape' – another example of turning to nature for comfort and solace without achieving any real change. In other words, a convenient distraction. This chapter hopes to attend to this issue, by drafting robotany as an intrinsic clause of a plant contract.

Robots, before Robotany

Before I focus on the recent relations between art and plants, as evidenced in the augmentation of plants with robotics, there are other kinds of robotany that should first be noted. An example of the robotic structures that incorporate plants is the bot created by Roxana Gherle, Sergiu Mezei and Mihai Garda-Popescu, where the plant is powering and powered by the bot, linked by a motivation to source water. These plant-bots are based on the IRobot Create platform combined with Lego NXT.[8] Linkoping University's Laboratory of Organic Electronics, in Sweden, led by Magnus Berggren, have

---

8 (http://www.ubbots.com).

built an electronic circuit using a rose. They used a conductive liquid polymer that was able to move through the rose, hardwiring the vascular system. By his own admission, they killed a lot of roses to get to that end.[9]

There are also robots that move your pot plants around the house to stay in the light (avoiding the shadows), which are called Inda-plants or faunaborgs, created at Rutgers University – where sensors alert the bots to shadow. Exactly like plants, these little faunaborgs shy away from the shade, appearing to move towards the light.[10]

Another example of non-aesthetic, industry-facing robotany are the robotic systems created to grow like plants.[11] The robot built by Barbara Mazzolai at the Italian Institute of Technology, Genoa, mimics the behaviour of real plants. This system has roots like real plants and, just like the real ones, it knows if or when the water, temperature or PH are off balance. The roots of these robot plants grow underground, and they do so by prioritising the environmental stimuli. Mazzolai's team first had to create a mechanism to mimic digging into the soil, as roots do. Nature adds material as it grows. The researchers fed the robot these artificial filaments so it could elongate itself to penetrate the soil. Real roots learn to bend away from heavy metals and other roots. The mimicry, in Mazzolai's robot, is done through electrically charged fluids. Overlaying more (or less) artificial filaments makes the robot root bend one way or another. Sensors are used to decode and read information about the environment and external objects. They know when they are touched and being subjected to forces.

In terms of applications, Mazzolai's 'plantoid'[12] can be used in spatial exploration or for preparatory exploration of potential mines underground. There is also talk of bolder applications such as in brain surgery.[13] Implicit in the concept of robotany – of art, plants and robotics (technology) – is that humans are infinitely capable of 'improving' nature. We can build it. We have the technology. We have the capability. Better, stronger, faster. Does any of that sound familiar? Well, it was the introductory by-line of the TV show 'The Six

---

9   Funnell, Antony. 'The Underestimated Power of Plants,' *Future Tense*, ABC Radio National 8 March 2016, http://www.abc.net.au/radionational/programs/futuretense/the -underestimated-power-of-plants/7227008.

10   (http://www.dailymail.co.uk/sciencetech/article-2321914/Plant-bot-worlds-robot-turns -household-plants-light-seeking-drones.html).

11   (http://www.sciencegymnasium.com/2013/08/robot-plants-grow-just-like-your-real .html).

12   (http://www.plantoidproject.eu/).

13   (https://ec.europa.eu/programmes/horizon2020/en/news/barbara-mazzolai -plantoid-project-robot-grows-plant-roots).

Million Dollar Man,' (1973) starring Lee Majors, which followed the story of a robotically regenerated astronaut who suffered an aero-nautical accident. If it's ethical for humans to re-build humans (the gamut of medical augmentation, not just fictional TV shows), then is it also ethical to re-build threatened plant life?

The Nomadic Plant

There are people who are developing new ideas about how plants grow and how to technologically support the plants that support us. These groups seem to be asking the question: what if the sun gets too hot to keep gardens and grow produce? What happens if there is less and less water available and poorer quality soil due to over-farming, and how can we solve the problem of alternative power sources?

Artist Gilberto Esparza has worked with robots for several years, attending to just this issue. In Mexico City, he colonised the city with urban parasites; these were robots designed to move around and search the city for energy. More recently, vegetation became his new passion, once he started to think about a symbiosis between plant and robot. His robot plant hybrid, the *Nomadic Plant*, responds to dryness. So when the bacteria or vegetation incorporated into the little mobile robot requires water, it moves to the closest supply of water, even contaminated sources, to suck up water to feed its plant. The poor quality water is filtered, or rather the bad microbial cells are decomposed, and turned into energy. Then there's more power to move the plant when it next needs water. This robot enables plants to survive in an environment that humans have damaged.

Esparza's creation is designed to be a multi-legged, autonomous robot powered by the microbial fuel cell. The spidery legs creep across the ground towards water. Its movement is a curious point for those of us interested in plant philosophy or nature philosophy. Two of the key reasons that plants have long been relegated to the background by humans as less relevant are their inability to move or think. This robot solves the immobility issue by ambulating on behalf of the plant. It's important to remember, though, that plants have their own particular mobility. Seeds travel many kilometres in the wind. Roots spread through the soil for miles and miles. There is constant movement within the frame of the plants: water and nutrients move through the vascular systems, photosynthesis making an active movement from sun to green, in one admirable process. While it is acceptable to say that plants have specific non-human movement, it is still interesting to see how a robot might solve the existential problem that humans have always attributed to the unwitting plant: immobility.

So how does the *Nomadic Plant* access the water? The sensors alert the bot to move towards water. The robotic arm sucks water through a tube and converts it to energy (electricity), with part of the water feeding the onboard plants. The robot's clumsy movements create a sense of sympathy.

The Nature Dome as Function and as Art

There are structures where the growth and mobility of plants is confined but within cathedral proportions. In Cornwall, UK, there is a massive geodesic greenhouse centre called the Eden Project, where thousands of rainforest trees and plants exist in an unlikely part of the world. There is a Perfume Garden and a Mediterranean Biome in the Eden Project too. These domed spaces – wherein the air is regulated and the humidity monitored – are experiments in futuristic ecosystems and they are popping up elsewhere in the world. As expanded and domed greenhouses, these entities are a curiosity – they are artificial vegetal spaces where the ecologies flourish but are not quite real. The Eden Project is a future-forecasting of creative ways to return to biodiversity via technology, artificiality and protection. These are extreme responses to the planet's state.

The glass dome or geodesic plastic structure is an interesting choice for a future enclosed world. Surely the sun's increased heat and dangerous rays mean that a different material should be used for our futuristic abodes? A plant contract, as presented consistently in this book, means we must remember to care for plants as much as the human in an imagined post-apocalyptic future. If there are underwater bacteria robots being developed to create robot/algae colonies, then we must approach these tasks with an ethical demeanour.[14]

Geodesic domes are eco-creative endeavours, vessels for the enactment of forms of human life mirrored in or by sci fi. The sealed dome structure appears in the 2013 US TV drama *Under the Dome*, plus *The Hunger Games* and *Silent Running*, among other narratives. These fictive domes are political hothouses, and for philosopher Peter Sloterdijk are potentially the future home of the human species as well. For Sloterdijk, domes contain ecologies of both biotic life and also future forms of governance. Shut off from the outside, strictly monitored and controlled, the 'human park' of Sloterdijk is a speculative space but one prefigured by existing arrangements for the plant world.

---

14   https://pdfs.semanticscholar.org/4be8/0226c4788229cb06950483db41305cd90ca7.pdf.

If we are living in times haunted by the end of life (extinction rather than death) and in constant dread of extinction, then this is clearly articulated in Hayden Fowler's cautionary artwork, *Dark Ecology* 2016. His work was a geodesic futuristic structure in clinical white, an oversized igloo that sat outside the entrance of the Museum of Contemporary Art, Sydney, and beckoned the viewer inside to view a *mise en scène* stage of apocalyptic fear. Fowler says, 'The "off kilter" is a way of expressing ideas of pervasive 'sickness' that's endemic to the Anthropocene (acknowledging as something that has been developing for millennia) I tend to layer a lot of information and ideas in the material construction of my work.'[15]

Inside Fowler's geodesic dome, there was a fallen dead tree and a creek overcome with sediment water, more like concrete than clay slip. Behind the dead tree was a cave where many antlers were piled up in a pyre, with not a single biotic being in sight. This was the after-life. This was future dystopia. He explained: 'I try and manipulate beauty a lot, I think it's an incredible tool for drawing people in, beauty is what we want to see and what we're drawn to see. But, then when you look at the work a bit longer, the 'sickness' begins to leach out. So in one image that longing for beauty, or exhilaration of momentary freedom, is then undermined by loss, death, sickness etc. Hope and despair in the same frame somehow.'[16]

Theorist Timothy Morton has been working with the term 'dark ecology,' the title of Fowler's artwork for many years now. It is the title of Morton's most recent 2016 book, which bears the sub-title 'For a Logic of Future Co-existence.'[17] Morton consistently presents the idea as an extension of an 'ecology without nature.' This is a strident reaction against Romantic notions of nature as Terra Mater (mother nature), of nature as background entity or as backdrop to human life. In other words, humans might not be thinking of nature. Nature might be thinking of us. Humans have been very preoccupied with our own importance and exceptionalism, and Morton suggests this constitutes a grim mistake. Nature is a construct of our minds, ecology is the real thing. Fowler represented this distinction in his anti-Romantic geodesic dome of decay.

Fowler says,

> Again this is partly a way of eliminating concepts of past and future in time, expressing or experiencing history as a single entity. So, directly

---

15   Gibson, Prudence. Interview with artist Hayden Fowler, 3 January 2017.

16   Gibson, Prudence. Interview with artist Hayden Fowler, 3 January 2017.

17   Timothy Morton, *Dark Ecology: For a Logic of Future Co-existence,* New York: Columbia University Press, 2016.

connecting the classical period to now as a means of drawing the continuum, connections, cause and effect. Also, I see myself as a Romantic artist (another largely discredited/marginalised movement) – as part of a modern movement that rose in direct response to industrialism and capitalism, and thus with the beginnings of climate change with the industrialisation of the environment, but also importantly, the industrialisation of humanity. Romanticism has been an ongoing voice of opposition. However, I want to re-contextualise within the present, which for me means incorporating ideas and aesthetics of the new knowledge of natural history, technology, filmic representations of apocalypse, dystopia, future utopian visions etc. It's this new combination of aesthetic elements which I think results in the 'off-kilter.'[18]

Fowler explained to me that he wants to grapple with the horrific path that humanity is on. He believes that the beauty in humanity has become marginalised and discredited: 'I see earth history as a single continuum, and understand humanity in the context of the past and future. By drawing in fossils of humanity from both the past and near future, I'm trying to understand or contextualise us in this continuum. As well as express the certainty of our end, our finitude.'[19]

In Fowler's related artwork, *Anthropocene* 2011, at Carriageworks, Sydney, he created another igloo form, raised on a multi-legged stand, with other spherical white shapes arranged around the igloo cave's front garden. Grass grew high. A video screen on a stand showed the activity inside the igloo. A murky milk-pond and a bromeliad were also part of the landscaping. In this performative work, Fowler 'lived' in the igloo. The audience could see him resting or sleeping inside his habitat:

> The design of work was very much based on Edward O. Wilson's ideas on Biophillia and research into the optimum environments that human's desire – shelter on a raised hill, overlooking water and open grasslands. So in a sense I developed a type of zoo exhibit or living museum diorama designed for the basic desires of the human subject. There were also aspects of depleted environment, surviving rainforest plants dotted amongst the new monocultured grasslands. Aspects of the post-apocalyptic survival fantasy, eating from a cache of tinned food, the colony of white lab rats that also lived on the space, the increasingly

18   Gibson, Prudence. Interview with artist Hayden Fowler, 3 January 2017.
19   Gibson, Prudence. Interview with artist Hayden Fowler, 3 January 2017.

poisoned water, etc., that all undermined the utopian aesthetics of the work. I guess the revealed sub-structure also pointed to the artificiality and precariousness of the human environmental creation. The loneliness of the single human subject also radiated the type of sadness and pathos that we know from many zoo experiences.[20]

The lonely last human. The beginning of early caveman life and ... the end too. Fowler has highlighted the plant element of a future dystopia. He has created a misfire between beauty and the grotesque. Mediated as art, Fowler not only contributes to the plant-technology hybrid discourse but also to the question of aesthetic effect. In his own words, beauty is what draws people in. A spherical form appeals to the human eye. So he has mobilised the vegetal, the futuristic, but it is based in an aesthetic form that is ultimately attracting.

The Underwater Garden

In terms of plant ethics and associated agricultural ethics within a discourse of a plant contract, there are some circumstances where technology and plant research are being harnessed to potentially improve upon nature. Such arrogance! This must be qualified: they are improvements to terrible conditions caused by humans during the Anthropocene. So even though some of these projects are 'improving nature,' they are improving a nature that has been damaged by human hands. It is no more than bringing the pendulum back to centre.

While researching these robotic plant projects, I came across some people who were creating miniature worlds beneath the sea. After all, there under the sea, is a massive resource that is shielded from the sun, kept cooler than land by its mass. Water temperatures have risen, it's true, but it is still cooler down there than out in the sun on land. The physics of waves and surging tides have yet to be utilised. It's an element of nature just waiting to be mobilised.

In the meantime, the entrepreneurs behind *The Nemo Garden* are starting that story in a small way. In 2012 the Ocean Reef Group, led by Sergio Gamberini, built underwater ballooned greenhouses off Noli in Italy. He wanted to make use of the properties of large bodies of water to grow crops, using the consistent temperatures and the natural evaporation of a surface of water adjacent to air.[21] The underwater garden fulfills this useful and resourceful technology where

---

20   Gibson, Prudence. Interview with artist Hayden Fowler, 3 January 2017.

21   http://www.nemosgarden.com/about-us/.

the latest system is aiding production and experimentation in the processes of agriculture, with an aligned conceptual appeal. While these examples are not participants in the art aesthetic module of this chapter, they are worth noticing in a wider context of plants and technology that are forming a new aesthetic. *The Nemo Garden* differs from the *Nomadic Plant* in the sense that it is fixed, as many plants are, in its place. There is no robotic ambulation but it exists in a deeply watery way.

Gamberini built transparent balloon structures, anchored them and filled them with air. *The Nemo Gardens* were built and the first live-stream footage of data began. Basil crops were grown hydroponically and an alternate light-footprint agriculture operation began. Sadly, rough seas and violent storms destroyed several of the greenhouses. A number of efforts failed. Now the Ocean Reef Group has rigid acrylic biospheres which are more hardy. Using remote access, they can operate the fans and see how the crops are going. The latest *Nemo Garden* (after the first failed attempts) is big enough for a couple of divers to enter and work inside.

These balloons are often used by other pharmaceutical research groups who are also interested in the concept of underwater greenhouses, as the way plants grow in these circumstances is of importance to their research. As Gamberini said, 'we moved to all rigid acrylic biospheres, we wired all bios rather than just one with multiple sensors; we created a complex system of remote accessible and activated mechanisms such as internal and external cameras and wi-fi is in all of our biospheres.'[22]

The suggestion is that these biospheres could have properties and qualities that allow survival that may no longer be possible on land, at some stage in the future. If the sun becomes too hot and the land too eroded, it is under the sea that we have options. These underwater structures could potentially dot the deep sea landscape if our old methods of on-land agriculture become untenable.

The Artists' Robotany: The Art-Plant Conundrum

Whilst there is a growing number of artists around the world who investigate the sonic and eco-transmissive nature of plants and trees, it is still an emerging phenomenon.[23] Understandably the critical analysis surrounding the science,

---

22   http://www.nemosgarden.com/about-us/.

23   Kac, Eduardo. 'Foundation and Development of Robotic Art' *Art Journal*, Vol 56, No. 3, Fall 1997.

on the one hand, and the artistic expression of the aesthetics of this work, on the other hand, lags behind the making. The politics of environmental art do not confine these artworks. The politics are part of the process of growth and loss, decay and regeneration. The desire to connect with nature and create artworks that are perpetuated by natural production is political. Art and environmental politics emerge together in most of the plant artwork discussed in this book.

Replaying the sounds of nature, mediated by synthesising software on a computer, can sometimes be a problematic process as it slips back into conventional aesthetic acts of mimesis, which do not function as well in the plant studies context. Artists such as Vincent Wozniak-O'Connor,[24] compost sonic artist Martin Howse,[25] Mileece,[26] Pia Van Gelder[27] et al. are creating 'sounds' of plant life but their mediation (via various software and syntheses) transforms the sound we hear. The plant sound becomes artificial in terms of science, but generative in terms of art.

The unusual and diverse sonic recordings of nature, created by various artists, open up discourse surrounding the concept of the communications among plants and our human limitations in adequately attuning to the frequencies that plants emit. We need the mediating technologies to overcome our inability to notice the hormonal, chemical and gas emissions which function as communication between plants and trees. The mediation constitutes the shift from scientific curiosity into an aesthetic realm. However, drawing on the comments made by Hayden Fowler, this kind of art relies upon a strangeness. Some of the sounds exhibited are not 'pretty' and are instead quite harsh to listen to. A droning hum. An uncomfortable whistle. What we hear is not the real sound of the plants but an artist's interpretation of those sounds. If those interpreted sounds are not aesthetically pleasing nor quite disturbing enough, can we count them as art or should we leave them as science? Plant artists are keen to experiment and explore the possibilities of language exchange between humans and plants. This motivation is a common link among artists working in this field of Critical Plant Studies.[28]

---

24   Vincent Wozniak O'Connor, *Runway* 2015. http://runway.org.au/category/contributor/vincent-wozniak-oconnor/.

25   Martin Howse, artist web site. http://www.1010.co.uk/org/.

26   Mileece artist web site. http://planet.mu/artists/mileece/.

27   Pia Van Gelder has done work with sensor monitoring the emissions of trees in the Australian outback and also created 'Psychic Synth' for Performance Space, Carriageworks, 2014, which experimented with human/non-human engagements.

28   Database held with writer.

The exhibition display of the 'processes' of sonic recordings of plant life often incorporates, as a major element of the work, the aesthetic experience. In other words, Bioart incorporates the making and creating process, the science of the making, the mechanistic structures of it, the biological conditions of its creation, the medical equipment used for it, and so forth. This becomes an aesthetic of pseudo-science within an art domain. This is a process of bringing science into the exhibition space as a means of displaying process. The question I am asking is whether this might constitute aesthetic criteria. What are the grounds for critically engaged, conceptually innovative and technically sophisticated artworks in critical plant contexts? Based on the artworks discussed so far, many of them consider the sensory values of the audience.

Robotany in the Bush

Sensorial experience is more relevant than ever in the art domain, due to its immediacy and reciprocity. Robotany, as I have defined it, refers to all these examples of plant-technology interaction, where systems are mutually beneficial. It comprises robotic elements incorporating plants, interactive computer plant growing, robotic systems inspired by the machinations of plants and various technologies that work alongside plants to improve agriculture, self-maintenance and sustainability. The artists who are using technology to communicate the sounds of plants, to create a closer aesthetic co-species relationship with plants, and who are engaging with computer software and plant life are doing so with an awareness of critical environmental issues.

There are several artists who are enacting the kind of natural contract that Serres presented in 1995 and since. For instance, Andrew Belletty has a long history as a sound artist for film and design. His artworks are more specifically an effort to experience nature beyond colonial settler conventions of seeing. It is sound and rhythm and reverberation that calls him back to the land and us back to a true experience. Belletty has been undertaking work in the desert areas to record the reverberations of the elements of his land. This work is more than a recording of nature, it is a fundamental and elemental soundscape, an acknowledgement that experience of the land is more than one sense – the auditory function of the land is merely one element. Here, Belletty talks about his work:

> I have been doing my practice as a cinematic sound designer with Aboriginal people, on Aboriginal country for 30 years. I use the term country in the same way that an Aboriginal person would use the word,

that is to describe the land as a conceptual thing that extends beyond a Western corporeal perception of the human and non human elements of the country.

My practice as a sound designer and artist makes it difficult for me to attenuate the sensitivities of my sense organs so I can listen to country in a non-human way. My default mode of perceiving country is anthropomorphic *and* Western, that is I see with my eyes, hear with my ears and touch with my body. For me to understand how this particularly vibrant tree is expressing and communicating with country I also need to take off my shoes and bury my feet in the sand, sit down in the sand, stamp on the earth with my hands and my feet and connect to the country bodily.

So I sat as the tree did, directly rooted in the sand, and as I did I could feel a rhythmic thumping, travelling up through my hips and thighs and feet. The vibrations were not audible to my ears, but I could feel them distinctly and strongly through my body, the vibrations resonating from a group of dancers a hundred meters away. The low frequency vibrations were carried easily by the loose white sand. The particularly vibrant tree a *Desert Grevillea* sat in the distance, a flurry of small birds, honey eaters, drinking from the blossoms. But when the dancers changed paths, the birds suddenly disappeared. I needed to inscribe this interaction between the tree, the sand, the birds and the wind to use in my artworks. So I set up apparatus that could detect and inscribe low frequency vibrations, placing ultra sensitive microphones, hydrophones and transducers at the base of the tree, near its branches and under the soil. Using these non-traditional techniques and technologies I was able to inscribe data that suggested a *tree's* point of audition rather than a human's.

I experimented with my own footsteps fifty meters away from the tree, and again the particularly vibrant tree fell silent. When I listened back to the recordings made from the *tree's* point of audition, I observed that the tree was acting as an antenna, transmitting and receiving energy between the loose white sand, the birds and the air. I could *feel* the subtle vibrations made by the bird's activity on the branches, the creaking of the trunk in the breeze and the straining of the roots in the loose sand. My attempt to embody what the tree was hearing, feeling and expressing in this sacred country, I started to get a sense that the tree was vibrant in ways that I was only just beginning to understand.[29]

---

29   Belletty, Andrew. *The Covert Plant*, Punctum 2017.

So the concept of robotany, as explored by Belletty in his work as a symbiosis of nature and sound microphonics, is a holistic approach to nature. The rhythms and sounds, movements and underlying reverberations are an important element of plant life. Plants do not exist alone but within a vibrant ecosystem. Belletty introduces complex theoretical and moral issues surrounding how we perceive nature and how humans perceive themselves. This has the potential to have a great impact on how we imagine ourselves as part of nature.

We still have to ask ourselves whether we are imposing and layering anthropocentric desires upon plants and the natural world. In augmenting nature with probes and sensors stuck into trees or clipped onto leaves of plants in order to emit signals and to transcribe the sound or reverberations of plant life, are we really exploring 'a *tree's* point of audition rather than a human's' as Belletty suggests or simply exploring human curiosity? Belletty does not use invasive processes for his recordings, and he does not claim to be communicating with plants or merely translating natural elements that are already audible and therefore available to humans.

Belletty's ability to acknowledge all parts of any given ecological environment is the basis of his ethical approach. Why? A criterion for aesthetics in environmental work needs to include the experience of the audience as well as the experience of the making and its environment – these parts make the sum. It is a constant across the centuries that the way the public is changed by the sensations of the artwork is crucial. Jacques Ranciere's thesis of 'aesthesis' relates to a connection between art and life, rather a separation between the two. His thesis of an aesthetic regime, which he believes began during the French Revolution period, is a cross-fertilisation of art and politics. Ranciere says, 'This is not a matter of the 'reception' of works of art. Rather, it concerns the sensible fabric of experience within which they are produced.' Ranciere's regime of art might include the critical plant studies art works that are being discussed here. This is because he argues for the imperfect, the lousy. Art is not the individual fantasy but the perception, affection and thought that revolves around them.[30]

In Belletty's work the political change or sensory capacity lies in the way the artist establishes a first-hand experience of the vibratory rhythms of nature and later delivers them to an audience in the exhibition space. Rather than mimicking nature, Belletty allows nature to enact itself on him, as an immersed participant in the bush. When we go to the gallery space and feel the vibrations under our feet then the experience is shared. The audience has access to his

---

30   Jacques Ranciere *Aisthesis: Scenes from the Aesthetic Regime of Art,* London: Verso 2013,
      XII.

story, his research, his process but they also have access to the very vibrations he experienced in the bush and relayed them for public impact.

It is natural to argue that the impossibility of escaping our anthropocentric ways, being decidedly human, returns us to a position of curious interest in the way humans, plants and technology are correlated. This impossibility of seeing the world outside the human can inspire us to think more broadly, even to think as a plant. By creating a memory of the land, by re-creating a performative experience of the land, Belletty moves beyond mere mimicry – which Ranciere described as being made possible due to the harmony of poiesis and aesthesis.[31] Belletty is not adding and building – he is showing the divisions and losses, he is giving us his experience of shedding our individual selves so that we can operate with all elements around us. He does this by de-stabilising the new environment in the gallery space. The usual object-subject art experience is changed into an *umvelt* or environment experience. It is an expression of plant and earth vibrancy, rather than a removed representation of it.

If we put aside our misgivings about imposing on, constructing and representing nature for just a moment, then there is room to think about the technological implications of robotany. Yes, perhaps it is no more than human exceptionalism to develop models of robotics that help plants find light, water and nutrients. But conversely, in aiding the environment we are so dependent upon, these robotanical artworks and engineering works are studies in future relationships, future means of agriculture and better models for understanding our place in the world.

Moving Plant-Bot Art

Mobility is an important issue in plant philosophy, as has been discussed in previous chapters. Interesting, then, that several major international artists are creating works that deal with the movement of plants in an aesthetic context. The following two case studies introduce two different kinds of movement: robotic and hybrid human-plant.

At the 2016 Venice Biennale, French Pavilion artist Celeste Boursier-Mougenot created a kinetic forest. Boursier-Mougenot's artwork was intended as an experimental ecosystem (calling upon the memory of 18th century follies) that responded to the idea that nature is constantly changing and evolving. Big pine trees, with equally large root balls, were pulled from the earth and placed on moveable pallets that wheeled across the gallery floor. The sap on the

---

31   Jacques Ranciere *Aisthesis: Scenes from the Aesthetic Regime of Art,* London: Verso 2013, 11.

trunk of the Scotch pine tree triggered sensors that activated the low-voltage electrical current that moved the pallets. The gallery space also had a sound component of low rustling noises, taken from recordings of trees nearby the pavilion.[32] Boursier-Mougenot is a musician, a composer, who spent 1985–1994 composing for Pascal Rambert. He is known for his music work that recorded the movements of birds on the strings of electric guitars.[33]

This artwork raises a couple of important ethical questions that may or may not flourish in my plant contract context. First, the three trees were removed from the soil they had been flourishing in. They were mature trees and large in size. Was this stress on the trees worth it, for the delectation or education of the art-going public? There is a problematic irony in damaging trees to raise awareness for damage done to the environment. This type of work risks perpetrating the very issues it is protesting against. The trees moved slowly and there was a sense of ambience and calm in the footage of the exhibition as it unfolded. This work was a way to remind humans of the succour of the natural world – its smells, its activities, its perpetual growth. It also highlighted the interaction and interspecies relations of technology and plants. We could see, for instance, the possibilities of augmenting trees and giving them 'movement.'

In terms of aesthetic effect, the dramatics of seeing these ungrounded trees in a major exhibition space, along with the sounds of the recordings of root activity in the gallery space, created an enjoyable art experience. Audience members could be seen smiling and revelling in the experience but political and ethical questions were raised.

Plants as a New Aesthetic Category

The second case study that relates to plants and human movement is the work of Danish artist Laura Beloff (see figure 6). She created *A Unit* and *Evidence* 2012, part of her series *Living in a Techno-organic world*, where she created a strapping apparatus which ties a pot plant to the side of a human shoulder. Rather than augmenting plants, Beloff plays with the idea of enhancing humans with plant life. This is an aesthetic of mutual care. It is a response to her concerns about environmental damage and what she describes as 'her

---

32   (http://www.contemporaryartdaily.com/2015/05/venice-celeste-boursier-mougenot
      -at-the-french-pavilion/).

33   Catalogue, *Revolutions*, 56 Esposizione Internationale d'Arte, French Pavilion, Venice
      Biennale, 9 May–22November 2015.

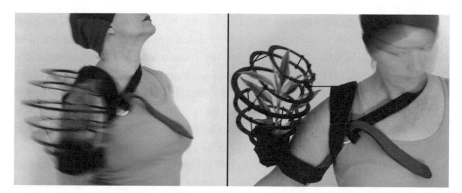

FIGURE 6   *Laura Beloff. A-Unit 2012. Pot plant, sensors, harness.*

reciprocal relationship with environment that is also getting increasingly modified with the advancements in science and technology.'[34]

Her interests were with how wearing vegetal matter in close proximity
could positively affect health and wellbeing. This project is an open scientific-
like investigation into co-species living but it has a dramatic or performative
element. Speculative and futuristic, her plant-wearing apparatus looks a little
like a sling used for a broken arm or else a pouch for carrying a baby. Both of
these types of harnesses are used (for broken bone and baby) as a support, a
preventative medical operative of care. So even the structure of *A Unit* ties into
the concept of an aesthetic of appreciation.

Academic Monika Bakke, who has written about Beloff's work, has spoken
about plants' embodied lives in artworks that incorporate heavy technological interventions. As part of a conference in Poland, she says that, 'Artists also
contribute to rediscovering plants often locating them in the technoscientif-
ic contexts but at the same times they are looking at the cultural history of
plants as well as into their *natureculture* futures. Art practices involving real
plants, both on a material and on a discursive level, open up a territory where
the complexity of plant life can be put forward and plants' ability to respond
can be emphasized.'[35] Bakke is interested in the way biotechnologies program
plants for our use. These subjectivities are an important part of critical plant
studies and particularly for artists working in this space, because development
of these issues sometimes comes at the cost of the very species we seek to
support.

---

34   Laura Beloff, artist web site, http://www.realitydisfunction.org/.
35   Conference talk, "What can art do for science?" WRO Center, 2015, (goo.gl/GkUrtm).

At a Helsinki conference on *Hybrid Matters*, Bakke and Laura Beloff both presented papers on plant subjectivities.[36] Beloff spoke about 'the survival of the prettiest' (which is obviously a pun on Darwin's 'survival of the fittest'). Beloff is talking about how audiences like an attractive artwork, even when the aesthetics are intended to be intrinsic to the scientific operations. Audiences feel saddened by bioart that does not have an aesthetic component. Is it enough to witness the technology and the biological creation of the work? This often means the 'art' component is no more than a record or archival footage – videos and photographs. Beloff refers to Jakob von Uekull's *umvelt* in the context of human perspectives and subjective survivals. von Uekhull's concepts of *umvelts* are developed as being interconnecting environments that can also exist within each other. The relationality of these environments and their interdependences are relevant today. His research on ticks and other animals and insects focused on their perceptions and the differences in experiences of all creatures of the very same environment.[37]

What Beloff's work reveals is an 'interesting' criteria of critical judgement. Beloff herself spoke about new aesthetic categories of art, referring to Sianne Ngai's 2012 new aesthetic category of 'cute and zany.' We could potentially add to this new category another new category – plants. What do plants add to the overall status of an artwork? Not so much cute and zany, but green and leafy. Can they be included in a litany of reasonable elements that aggregate to create a successful body of work in its entirety? This is an extension of the intersections of the *umvelts*. They do not exist independently of each other. Art is a natural beauty, according to Adorno, so could we use this point to reverse the emphasis that vegetal life could be a legitimate criterion for judging the aesthetic success of a work?[38]

Beloff says:

> *A Unit* references the understanding of Gregory Bateson that the unit of survival in the biological world is the organism plus its environment (Bateson 1969). In this work 'the unit' consists of a human plus a fragment of our natural environment – a plant. The work focuses on the

---

36   'Hybrid Matters' Sympoisum Theatre Academy of the University of the Arts Helsinki, Nov 25th, 2016. www. symposium.hybridmatters.net.

37   Jakob von Uexküll, "A Stroll Through the Worlds of Animals and Men: A Picture Book of Invisible Worlds," *Instinctive Behavior: The Development of a Modern Concept*, ed. and trans. Claire H. Schiller (New York: International Universities Press, Inc., 1957), pp. 5–80.

38   'Hybrid Matters' Sympoisum Theatre Academy of the University of the Arts Helsinki, Nov 25th, 2016. www. symposium.hybridmatters.net.

potentiality of beneficial impact of natural green environment for human health. In this case, the environment is constructed as a miniature
green area to be worn by an individual(s). *A Unit* speculates on the concept of green environment and its beneficial impact. It experiments with
an idea of wearable miniature green space that becomes part of one's
everyday existence and asks if this can be considered as natural environment with potential health benefits?

Additionally the work *A Unit* has another layer which speculates on the
future development of a human-nature relationship. When environment
changes it impacts the development of an organism and vice versa. The
work *A Unit* asks: when humans are modified and nature is manipulated
what kind of relation will form between them? This wearable device is
designed to be housing a GM-plant or other primarily human-constructed
plant. It is a training device for our changing relation with organic nature
for the future when both humans and nature are artificially modified or
constructed.[39]

This art work is a reflection upon natureculture, a means of elaborating old
efforts to observe and formalise natural elements, of endeavouring to search
for fundamental archetypes in nature. All these scientific endeavours of
German idealism, of the Enlightenment, persevere today in terms of a basic
human need to absorb the beauty of nature. Beloff takes these ideas further by
adding the technological enhancement. But who or what is Beloff enhancing?
Is she improving nature or the human? There is the implicit suggestion that
the plant, so close to the human, will improve well-being and health – that the
plant is improving or augmenting the human. The human is *improved* but the
plant has been *modified*. The plant attached to the wearer's shoulder is intimate, private – a little piece of greenery for that person alone.

Beloff says that the contraption works as a measurement and evaluation
source for a 'quantified self.' Her related work *Evidence* collects data of wearer's geo-location – stress levels of the wearer, that is – through pulse and skin
resistance. This data is collected and made available online. This is all done
to assess the qualitative effects of the immediate plant environment on the
health of the wearer as she or he moves. The device can also be used to measure the stress-levels related to usage and non-usage of *A Unit*.

Tuija Kokkonen is an artist who performed a work, *Chronopolitics with dog
and tree in Stanford*. With Alan Read, she worked with texts from the past where

---

39   Laura Beloff, artist web site, http://www.realitydisfunction.org/.

animals and trees had conversed e.g. Tolkien.[40] The idea, as she expressed it, was that we have transformed cohabitants (companions) into representations and products, as Derrida predicted. These creatures made sense of us and our reason to be. Humans now live in self-designed and self-constructed environments, only with ourselves. She intended that her performances would diversify that thinking.

Kokkonen's work with animals can be extended to concepts of plant art where both disciplines converse and make sense of each other. In these self-designed worlds that we endlessly propagate, is there any room left for other species: the vegetal? If so, how? Is art-as-mediator the only answer?

The Awkward Plant

Can an artist's playful work with artificial plant life add to this debate about art, plants, technology and aesthetics? Australian artist Tully Arnot's *Nervous Plant* comprises a pot plant (a fake one) placed in a standard office setting – that is, horrible white lighting and boring office furniture.[41] The fronds and trunks of Arnot's pot plant have 19 servo motors with sensors so light causes constant movement, programmed to move 10 degrees. Shadow stops movement.[42] The proximity sensors are activated when people come close. The artificial plant fronds stop moving when people are close, then starts its covert movements again.

Arnot here is interrogating sociable robotics. He was motivated by research into language and is interested in the moment when simulated life is close to the real. This pot plant, we must remember, has no learning or memory skills. It has no root system, no rhizomic operations guiding its actions. So as people wander past and the shy pot plant freezes with social anxiety, we might laugh a little but it does serve as a reminder of how humans, threatened by the climate change they have engendered, are the ones who have threatened other species for several centuries.

This artist is working with robotech movement to amplify our relationship with nature, rather than merely acting as a conduit between nature and the human. Arnot says, 'I'm not interested in the original Unheimlich, things like

---

40   (in A Journal of the Performing Arts, Vol 19 Issue 3, 2014).

41   Blamey, Peter. 'Unexplored Functions/Everyday Objects: Interview with Tully Arnot.' *Das Superpaper*, Issue 33, November 2014.

42   Ibid.

the animated doll ... I'm interested in something less human-based ... I'm interested in how an animated or simulated object can have the same kind of feeling as interacting with a robot or prosthetic form.'[43] These are issues in social robotics – not what does the robot think but what do we think about what the robot might think? Or in this case, what do we think about the idea that the fake plant might be thinking? Social robotics often attempt to deal with the humanistic elements of robots, but of course it is also gauging how humans are responding to the sociability of robots, or HRI (human robot interaction).

Like Beloff's shoulder/plant attachment that uses sensors to create new data about human/plant information, Arnot's work is compelling in a robo-tanical sense. There is a shift away from the human to a deeper sequence of questions. The human is still present in the aesthetic domain but the expe-riential perspective keeps swinging around from plant to human and back again.

Neither Beloff nor Arnot's artworks could be labelled 'Conceptual Art.' This is the category problem that plant artists face. They are not Conceptual artists, Postmodern artists nor even Bioartists. How can we categorise the plant artists? Performance? Installation? Environmental? If we conform to the concept of the plant contract expressed at the beginning of this book, then perhaps the desire to categorise this vein of art is not constructive. Better, in fact, to leave the overall categories for this kind of art out of the wider system and only con-sider the criteria that aggregates to make them in the first place.

The Signal Centre

Connected multiplicity is at the heart of plant-human-technology and is a means of suggesting economic and social togetherness, which is an intrin-sic element of a plant contract. This was literally expressed and borne out in Latvian artists Rasa Smite and Raisa Smits' human-plant communication project *Talk to Me* at the RIXC Centre for New Media Culture.[44] This work aimed to communicate the superior functioning of plants and the removal of 'the human' from its apical ontological position. Based within the scientific enquiry that talking to plants increases and enhances vegetal growth, Smite collaborated with others on this project to talk to plants.

---

43   Blamey, Peter. 'Unexplored Functions/Everyday Objects: Interview with Tully Arnot.' *Das Superpaper*, Issue 33, November 2014.

44   Smite, Rasa; Smits, Raitis; Ratniks, Martins. *Talk To Me: Exploring Human-Plant Communi-cation*, RIXC, Riga, 2014.

This project involved a number of pot-plant vines on trellises, in seven gallery spaces across Europe. Thousands of visitors were encouraged to communicate via software synthesis for speech or via email or text message. The messages sent signals to the plants via activated sensors. The plants provided a source of information, a system of complex networked and aggregated action. Photosynthesis, rhizomic activity, transfiguring neural properties and the lengthy process of growth extending beyond the lifespan of the human (especially in the case of trees) were elements or conditions that the artists discussed with visitors to stimulate a closer understanding between species. Rasa Smite's work generated the possibility of talking together and collaborating with biotic life, using abiotic computer interfaces.

In a similar vein, *Flora Robotica* is a four-year project (2015–2019) which is part of the Societies of Symbiotic Robot Plant Biohybrids, as Social Architectural Artifacts. Their work ties into communication and the intersection of plant life and technology. They 'investigate closely linked symbiotic relationships between robots and natural plants and to explore the potentials of a plant-robot society able to produce architectural artifacts and living space.'[45] Podcasts, symposium, consortiums, publications and exhibitions are part of this project.

Intelligent technology and 'intelligence' in plants are emerging in parallel to changes to our green environment. These groups are working to develop new ideas and solutions for greener cities and better growth. They want to get plants and robots working together, not only for human ends but to better understand plants as deeply sensitive organisms. Symbiosis, a meeting between plants and robots, is a process of species development that can work together. Computation and calculating is a mechanised skill that can support the plant: measuring how the plant is growing, how fast it grows, and how much water it evaporates. These measurements decide how robotics can help. Compatible robotic and plant systems is their task.

Flora Robotica is a project across four countries. There are teams of mechanical engineers, plant biologists, evolutionary plant scientists, swarm intelligence scientists and artificial life zoologists and mechanotronics – these are the skills for the project. This basic research project is rich in future applications, such as living room plants that can connect to the internet, and architecture that is self-repairing and self-growing. It is an attunement between plant/robot and its environment. This might be a case of colonising plant-robots.[46]

The futuristic and speculative sci-fi qualities of such projects as *Flora Robotica* were influenced by artists of the past. These include Tom Zahuranec's 1972 *Radio event 20: Rhododendron*. Electrode sensors fed information into

---

45   (http://www.florarobotica.eu/).
46   (https://www.youtube.com/watch?v=Byo55asQUwM).

the synthesiser and into software to make sound. Zahuranec saw the plants as mediators between the ESP of audiences and of plants. The extra sensory perception of plants (known as the Baxter Effect) was allegedly transferred to the plants via audience members' thoughts. A psychogalvanometer was used to measure the ESP of people. Bouncing sound in the speaker meant the plant was getting more elated from the feelings of people. Synthesiser sounds changed according to emotions of people.[47]

Recently, plants have been found to produce sound in the form of acoustic emissions in the range from 10–240 Hz, which falls within part of the human hearing range. Vincent Wozniak-O'Connor, an Australian sonic artist, has been experimenting with plant synthesis. He has been revising plant/human ecologies where we strain to hear their sounds. Ecological representation is shifting towards a communicatory duality rather than a removed aesthetic.[48]

Listening and hearing are mutually exclusive. Try as we might, there will always be a system of vegetal life that humans have no access to. We can only speculate on what plant sensibilities are. What is the connection between art and speculative aesthetics? Theories of inhuman networks and of 'vegetal existentialism' have been developed in Michael Marder's research. His inquiry seems not to ask 'can plants think?' but 'what do we think plants think?' This absorbs post-human consciousness into a different and ontologically-flattened view of life.

Recent biological discoveries have directed many writers and artists towards a trans-disciplinary approach to bio-art. An awareness of plants and insects as autonomous and agented things fuels a consciousness of human/ nature relationships. Artists Christa Sommerer and Laurent Mignonneau created an interesting artwork at the Project InterCommunication Center in Tokyo in 1994, entitled *Interactive Plant Growing: A-volve*.[49] These collaborating artists were rightly interested in the rate of growth as a theory of form and time. Rather than seeing a plant as an organism that exists in a static time

---

47 Zahuranec mentioned in: "Bio sensing art in the 1970s" *Datagarden*, 20 SDeptember 2011, http://www.datagarden.org/blog/richard-lowenberg-interview.

48 Other sonic artists comprise Mileece in LA 2013. She uses analog bio-emissions – current off electrode on leaf is conducted into amplifier and computer software takes data and animate sound. Sound is generative not a sample. Beautiful sound and visuals – Youtube. Examining both emergent and established approaches to site-based-sound, artists like Emptyset, Leslie Garcia, Christina Kubisch and Luke Jerram feature as complications to the development of sonic approaches. These artists demonstrate discrete practices for making generative sound and installations based on manipulating the connection between sound and site.

49 Sommerer, Christa and Mignonneau, Laurent. *Interactive Plant Growing: A-Volve*, Tokyo: InterCommunication Center, 1994. http://www.interface.ufg.ac.at/christa-laurent/WORKS/ FRAMES/FrameSet.html.

and space, plants need to be seen as constantly changing, moving and shifting: an ongoing event across time and space. As they say in their catalogue
essay, 'without movement evolution is not possible.' It is precisely because philosophers and scientists have not understood the rate and time-frame of plant
growth that they have pigeonholed the vegetal world into a non-movement
category.

What Sommerer and Mignonneau did in 1994 was to create an installation
where virtual plants were growing in three-dimensional space, as relating to
real plants growing in real space. Viewers were invited to come up and touch
the five real pot plants that sat in front of a large screen. By touching the plants
or moving their hands in front of the real plants, they could control and influence the virtual growth of the program-based plants. The artists maintained
that the real plants were responding to the emotions of the humans who
approached or touched them, saying, 'What is important here is the physiolog-
ical and emotional changes in the plants themselves.'[50]

There was some questionable science to this installation. The curator Toshiharu Ito explained how the audience could see the immediate and concomitant blooming of flowers and opening of leaves in the screen plant versions
when they moved their hands near the real pot plants. He said, 'As the viewer
puts his/her heart more and more into the hand movements, the growth of
the virtual plant accelerates, spreading across the screen in all directions.'[51]
The computer program (Silicon Graphics) was supposedly responding to the
changes of emotion in the viewers which then changed the 3D virtual environment. There is no software that picks up on emotions, only the symptoms of
certain emotions – heat and heartbeat, for instance. This, of course, is an issue
of language.

The artists and curator claim to be working with the 'emotional changes
of the plants themselves' and Toshiharu says. 'It's strange, but plants have
emotions in the same sense that we do. Plants can express preferences in
music as well as likes and dislikes regarding human beings.'[52] There is no
scientific research to date to support this statement. We have to be careful of
our language as artists and writers, as well as scientists, or we do more damage
than good.

---

50   Sommerer, Christa and Mignonneau, Laurent. *Interactive Plant Growing: A-Volve*, Tokyo: In-
      terCommunication Center, 1994. http://www.interface.ufg.ac.at/christa-laurent/WORKS/
      FRAMES/FrameSet.html.
51   Ibid.
52   Ibid.

These claims about the feelings of plants do not denigrate the value of the work. The five plants – ferns, cactus, vines, trees and mosses – all had sensors attached to their roots. The distance between the human's hands and the sensor of the roots decided the effects on the screen via electrical voltage differences. The signals received by the sensors were amplified, filtered and converted to create data values. Then these values were identified as variables, which changes the image on the screen.

Conclusion

It may be too soon to draw up a standard of criteria that places the focal point of art, plants and robotics within aesthetics. But it is the role of a *plant contract* to begin that kind of discourse. The conclusion so far is that plant art comprises an experience of the vegetal, an experience of the uncanny (in terms of the robotic uncanny valley, but also in terms of the uncanny relations that we have had with nature and will have with natural elements in the future). Criteria may also include an ability to create empathy or sympathy for plants, and creating a means of mediation between human and plant via technology or hybridity. If these artworks bring attention to and awareness of critical plant studies issues, then this could be added to the list of criteria for a new aesthetic form that participates in and represents *the plant contract*. The robotic as mediator between plant and human can finally be considered a criterion in this new field of art and plants.

Michael Marder discusses Aristotle's suggestion that plants have souls and yet their teleology is goal-less or indistinct. Without goals or souls they supposedly lack vitality. However, they constantly grow, they continue to develop and it is this ability to emerge and change without pause that creates desire. It is their difference and their adaptability that we desire to share. An irresistible temptation to share plants' ability to nourish and be nourished. The desire is to increase the interdependence. This could also be said of the process of artistic creativity: to grow and to have souls, but without any apparent distinction. Art does not need to heal or to lecture. It just needs to be created within a political and aesthetic realm. If the experience of art changes us, then it becomes increasingly desirable.

If metaphysics can be maintained in this new area of plant art study, then there is a place to create a formal means of symbiosis between nature, culture and aesthetics for the future. Aesthetics is an entire network of intrigues and discoveries, of calculations and thought. Plant art, by its own definition

(part plant and part art), is divided again by robotics and technology. Therefore, robotany becomes endlessly diminished and depleted, but at the same
rate as it multiplies. Plant art is a clear representative of this constant division/
depletion and regenerative growth. Much like the environment it refers to.

As such, plant art may have the capacity to mediate the division between
human and nature. If a plant contract comprises social, legal and scientific
elements, it might also entail a discourse of collectivities – commercial, agricultural, political and artistic. Science shows us the object's point of view. Plant
art shows us the plant's point of view.

CHAPTER 4

Bio Rights: Earth of Agonies and Eco-Punks

Nature Rights and Activism

This chapter canvasses constitutional amendments of nature rights that have been made in several countries to date. Conversely, it also allows the outlaws and the sometimes non-law abiding activists to be noted and counted. In the lacuna between the law and nature rights are those individuals – theorists, artists, writers – who are subverting and deactivating our extant perception of plants as less relevant than humans.

Nature rights, in particular the rights of plants to have rights, is a major issue in a plant contract discourse. There are individuals whose work I can honour here, that have activised the potential of natural life. Michel Serres' poetic theory that incorporates mythology and tales is one.[1] His writing is complemented by other kinds of persuasive writing in this field such as Michael Marder's and Luce Irigary's many books and Paco Calvo's 'The Philosophy of Plant Neurobiology: A Manifesto.'[2] Serres' natural contract is not a legal document because, clearly, plants can't sign any parchment paper. It is a moral obligation, that calls for ethical environmental care.

Serres' natural contract, and Rousseau's social contract before that, were manifestos or stories rather than legal agreements between two parties. In that vein, this plant contract is a story. It is a story of seeking to change our perceptions of nature, particularly plants. Most importantly, it is a story of how art and artists are the great mediators of this shift from non-agented to agented, immobile to locomotive, and passive to active (activists) in vegetal life. Artists have the capacity to grow at the same rate as plants – it is through the aesthetic beauty of plants, and the concomitant beauty of art that we can see beyond production, consumption and usage of plants.

As Rousseau said in his social contract, and here I am applying it to plants, the group will can achieve a great deal more than an individual will. The 'group will' is the 'creative will.' This is a kind of mass natural voice. The sovereign is nature. The sovereign is creativity. The activities are the laws, executed via action. To have a philosophy of law, a jurisprudence that focuses on the plant

---

1 Michel Serres, *Biogea*, Minneapolis: Univocal, 2012.
2 Paco Calvo, 'The Philosophy of Plant Neurobiology: A Manifesto,' *Synthese*, Vol 5, No 193, 2016: 1323–1343.

world, we need humans. Yes, we are bound by that limitation. But likewise, there is no earth jurisprudence without the plants. Can there be a constitutive change to the human law that makes sense for a vegetal world? *Wild Law's* Cormac Cullinan says that the ever-changing quality of nature, of a different species, needs to be addressed in a wild law.[3]

A plant contract cannot be situated where power alone sits with either the human or the plants. Instead, a plant contract must be one where the rights of all are acknowledged and grasping for economic power will incur a threat to this kind of resolution. Both parties will otherwise suffer. Could this be a Rousseauian means of 'voluntary agreement'? This is an impossible task, as we cannot consult the plants and ask them how they would like to be represented in a court of law. Instead, we can think like plants and note that their 'will' is to thrive and grow as part of an ecology. This we know. This we can represent.

Preserving the common interest is crucial to *the plant contract*, where the individual and the community are served at once. For me, the only action at hand is the force of art to present new perceptions of plant life to a broader audience, and the chance to communicate that transformation in words, in this text. Joanna Zylinska writes that the human/non-human discourse makes us think better.[4] All that we do, say, present, express and play will always beg the question: what if plants not only responded to me but began to direct my behaviour? The plant contract is global but undertaken at the individual human level. In this way, it enters the world. Via the artists. Via the eco-punks.

Eco-punks comprise the writers working for eco-justice, such as Cormac Cullinan. Eco-punks include artists playing with human neuroplasticity to trigger hybridity and trans-species perceptions. They are artists working with marine ecologists to regenerate sea vegetation, such as public work artists Turpin/Crawford, or using plants to highlight the failure of certain social, religious and political structures like the Church, such as installation artist Anna McMahon, both of whom will be discussed shortly. Of course, the most admirable eco-punks of all are the plants. Eco-punks, here, refers to humans-for-plants and plants themselves. This discussion of plant rights and eco-punk plant advocacy fits within a political plant contractual framework.

Michael Marder argues that the 2011 Occupy protest movement sits parallel to plants because of its self-replication in different spaces and its civil

---

3   Cormac Cullinan, *Wild Law: A Manifesto for Earth Justice*, Devon: Green Books, 2011.

4   Joanna Zylinska, 'Bioethics' in *Telemorphosis:Theory in the Era of Climate Change* Vol 1, Open Humanities Press, 2012.

disobedience.[5] Contemporary art, like the Occupy movement, is socially and civilly disobedient – a determined demonstration of agency and political force. The Occupy movement protested against social and economic injustice and inequality. Marder explains that Occupy was a politics of space, a physical occupation of various places where the activist occupants were not welcome, such as banks, corporate foyers and university campuses. The Occupy protestors were the weeds, the alien beings, threatening the organization and structural order of our consumer-driven world.

In this vein, art and plants are expressed in this book as more than an aesthetic, but also as legal, social and political entities. While plants cannot defend themselves within a human context, they can defend themselves outside the human. Marder says, 'Purely vegetal beings do not protest, do not set themselves against anything, do not negate – symbolically or otherwise – what is. But if we act as though we were them, following a useful theoretical and practical fiction grounded in the vegetal heritage of the human, we would need to follow a non-possessive, non-appropriative way of being, resonating, at once, with the conclusions of botany and with the image of postmetaphysical ethical subjectivity.'[6] Marder reminds us of the physical corporeality of plant existence, the subjectivities of plants and the hubris of human ownership of plants.

Likewise, art in the 21st century is almost always a protest, whether emotional, aesthetic or social. It is an expression of artistic means that has a political entelechy, by virtue of the making of art at a particular time and in a particular place. Artists have the right to make art. Marder reminds us there is room to consider the vegetal world's right to have rights. Whilst there has been a convention of considering plants as passive, this concept has now changed due to new discoveries in science that have fueled Marder's philosophical work. He refers to the animal movement's work on the right to have rights, based on philosopher Hannah Arendt's notion that rights-bearers must be citizens, and that citizens are defined as being entangled with sentience and able to feel pain.[7] Now that we have a deeper understanding of plant sentience,[8] it is timely to develop that progression from animal rights (and a natural contract) to plants rights (and the potential for a plant contract). A plant contract

---

5   Michael Marder, 'Resist Like a plant! On the Vegetal Life of Political Movements,' *Peace Studies Journal*, Vol 5, Issue 1, 2012, 25.

6   Ibid.

7   Michael Marder, 'Should Plants Have Rights?' *TPM* 3rd Quarter 2013.

8   Michael Marder, 'Do Plants have their own form of consciousness?' *Al Jazeera*, 25 June 2012, http://www.aljazeera.com/indepth/opinion/2012/06/2012619133418135390.html.

considers plants' rights to have rights. In this chapter, the plant contract will continue to be embedded in art praxes – the making of it and the writing about it – as a mediation and an expression.

Ethical Dilemmas

Before delving deeply into the rights of plants, however, there are plant ethics frameworks to consider. Matthew Hall explains that plants 'generate the conditions of their own flourishing.'[9] We, the writers, could learn much from this particular plant skill. Humans are sometimes inhibited by structure and order. And yet we – writers and all the other strange humans – do many things without thought and care, without generating the conditions for flourishing. As a species that increasingly relies on plants in our diets (with rising numbers of vegetarians and vegans), have we thought enough about what kind of impact that has on the flourishing of natural plant life? When I put a delicious forkful of English spinach in my mouth, dripping with vinaigrette, must I feel bad? Should I feel sheepish because I haven't taken into account the 'dignity of plants'?[10] If plants can sense reverberations of caterpillars and retract their leaves when under threat; if they emit chemicals when dangerous species overeat them and communicate to one another if there is imminent danger – in other words if they exhibit cognitive-like behavior – ... should we eat them?[11] It is a tempting and volatile discussion point, but to feel guilty about eating a salad or popping a baby tomato in your mouth does not really address the true issues of vegetal being. It obscures them from their own being, by imposing an anthropocentric morality.

In addition to the idea that human morality may obscure plant ethics, perhaps the vegetal state of being defies an application of the law. There has been little progress in earth jurisprudence, especially in the context of nature rights. It comes back to the old chestnut that if humans don't own it, they don't care for it: wild tracts of land don't fit neatly into systems created by humans, for humans. The inner functioning of plants, their operations, their sensory

---

9   Matthew Hall, 'Plant Autonomy and Human-Plant Ethics,' *Environmental Ethics*, Vol 31 Summer 2009, 169.

10   Florianne Koechlin, 'The Dignity of Plants' *Plant Signaling and Behaviour*, Vol 4, No 1, 2009.

11   Richard Firn, 'Plant Intelligence: An Alternative Point of View,' *Annals of Botany* 93, 2004, 345–351; Anthony Trewavas, 'Aspects of Plant Intelligence: An Answer to Firn,' *Annals of Botany*, 2004, 353–357.

awareness of their environments and capacity to thrive don't fit into our human legal system either.

The lack of legal representation of plants (although the Buddhist ordainment of trees has an ex-legal power of protection) means that it is an emerging field.[12] It is a field that is too loose and unwieldy to be limited to precise regulations and laws as yet ... so constitutional changes still need to be drafted in most countries. It will be a tricky process to refine and harness. The discussions in this chapter acknowledge the successes so far – such as the Swiss and Ecuadorian legislation amendments in 2008, governments that added a documented clause to establish the rights of nature.

In 2014 the New Zealand Wanganui River established an ecological area as a legal entity that can be represented in a court of law. The 2014 Te Urewera Act granted an 821-square-mile forest the legal status of a person. The forest is sacred to the Tūhoe people, an indigenous group of the Maori. For them, Te Urewera is an ancient and ancestral homeland that breathes life into their culture. The forest is also a living ancestor. The Te Urewera Act concludes that 'Te Urewera has an identity in and of itself' and thus must be its own entity with 'all the rights, powers, duties, and liabilities of a legal person.' Te Urewera holds the title to itself.[13]

These are earth jurisprudence matters, where the concepts of law precede legal courtroom precedents. The International Rights of Nature Tribunal, held in Ecuador and Peru in 2014 and Paris in 2015, are all charted by the Global Alliance for the Rights of Nature.[14] Declarations and proposals are starting to proliferate, such as the International Union for the Conservation of Nature who are working towards a Universal Declaration for the Rights of Nature.[15] However, high profile successful cases, at the level of human rights cases such as the Australian Mabo decision, have not filtered through the legal system yet. There is still a large amount of grassroots change to take place at the level of the public psyche. First must come the perception shift, second must come the consensus, and finally the changes to the governance system.

The main problem remains. Exacting an ethics or moral code for a renewed relationship with plants will be difficult because of the issue of consumption.

---

12   Susan Darlington, *The Ordination of a Tree: The Buddhist Ecology Movement in Thailand*, Albany: SUNY Press, 2012.

13   http://www.legislation.govt.nz/act/public/2014/0051/latest/DLM6183601.html. Accessed 6 January 2017.

14   http://therightsofnature.org/ Accessed 6 January 2017.

15   http://therightsofnature.org/iucn-adoption-of-2020-universal-declaration-of-the-rights -of-nature/ Accessed 6 January 2017.

It is an important topic, because moral codes are pitted against moral codes. The Animal Movement – groundswell movement against bad agricultural and animal husbandry practice, and the endemic negative effects on the environment of such industries as cattle production – means that a growing percentage of the population is vegetarian or vegan. This moral code, however, relies upon humans being able to eat vegetables or vegetal byproducts instead.

This is a confronting philosophical dilemma for those who have followed the trajectory of abstinence from all products that harm animals and insects, that is, veganism. Perhaps now is the time to reassess our food consumption. Rather than abstain from given food groups, could this be the time to consider eating all things in moderation and with a moral approach? Yes, go ahead and sanction food that shows no moral care in its industry: issues of humane slaughter, responsible packaging, appropriate pricing. Sanction products that do not follow ethical treatment. However, giving up all food groups except for plants is probably not the best solution any more, given what we know about plant behavior, sentience and non-cognitive intelligence.

The Art Dilemma

As for the ethical dimension of the intersection of art and plants, which is the main operation of this book, arts and culture writer Ellen Mara De Wachter maintains that plants 'manifest the same energetic properties that govern all life. As such they have been aestheticized through a full spectrum of expressive possibilities – from the purely ornamental or festive to the deeply symbolic, ritualistic and devotional.'[16] She brings to our attention that, like the incorporation of humans in performance art, many artists are recruiting live plants in their works. She cites Mexican artist Abraham Cruzvillegas, Brazilian artist Tonica Lemos Auad, US artist Rachael Champion, UK duo Cooking Sections and Rachel Pimm. Wachter's point is that plant-art moves our attention away from what she calls the 'propaganda of recycling and hybrid cars' and towards a consideration of other beings as if they really mattered.[17] These are eco-critical arguments in the sense that they bring ecological awareness and consciousness to the realm of art and aesthetics.[18]

---

16   Ellen Mara De Wachter, Ellen 'Art and Life,' *Frieze*, 24 April 2016.
17   Ibid, p. 8.
18   Serpil Opperman, 'Theorizing Ecocriticism: Toward a Postmodern Ecocritical Practice,' *Interdisciplinary Studies in Literature and Environment*, Vol 13, No 2, Summer 2006.

Bryan Bannon investigates the role of aesthetics in environmental ethics by pointing out that aesthetic and ecological evaluations are not always aligned.[19] This is helpful in a process of making sense of historical and future expressions of art as they intersect with nature, especially plants. Without conventions of beauty or truth as an ordering or organising principle, it becomes more difficult to categorise new bodies of work that incorporate plants but are not working as isolated aesthetic objects. They present more than a beautiful (or truthful) image and they also reference the natural world in a political way.

Glenn Parsons and Allen Carlson are called on by Bannon for their comments of how land art is an affront on nature. They consecutively refer to interventions by artists as an alteration of the beauty that was already there and also of the aesthetic appreciation of that beauty, changing nature to art and thereby obscuring nature.[20] However, Bannon counterpoints that artists such as US land artist Andy Goldsworthy, who leaves carpets in forests or arranges stones in aesthetically pleasing ways but in a disconnected context, creates 'the discomfiting in nature' and therefore accentuates 'new possibilities for collaborating with nature.'[21] There are dangers of using the expression 'collaborating with nature,' however, as it assumes that nature can participate in the mutual collaboration.

Small Misdemeanours

On a personal level, I no longer see plants as a backdrop or a background setting for human activity. I've begun to notice exactly when my mulberry tree loses its leaves and how quickly it seems to burst with new growth overnight in early September, as though on cue. I've begun to pay attention to the way my neighbours garden, or cut flowers for their table, when they fertilise and how vigorously they prune. I am conscious of writing in a plant-like way.[22] I even

---

19   Bryan E. Bannon, 'Re-envisioning Nature: The Role of Aesthetics in Environmental Ethics,' *Environmental Ethics*, Vol 33, Winter 2011, 415.

20   Bryan E. Bannon, 'Re-envisioning Nature: The Role of Aesthetics in Environmental Ethics,' *Environmental Ethics*, Vol 33, Winter 2011, 432; Glenn Parsons, *Aesthetics and Nature*, New York: Continuum, 2008; Allen Carlson, 'Is Environmental Art an Aesthetic Affront to Nature?' *Canadian Journal of Philosophy*, Vol 16, 1986.

21   Bryan E. Bannon, 'Re-envisioning Nature: The Role of Aesthetics in Environmental Ethics,' *Environmental Ethics*, Vol 33, Winter 2011, 433.

22   Prudence Gibson, 'Climate Cry' and 'The Underground Garden,' *Climate Century*, Vital-statistix, Adelaide. 2015.

asked a friend to teach me how to graft plants when I was staying down in the Kangaroo Valley.

Matt is a nursery owner in the valley who grafts three varieties of plants. To learn the technique of grafting, we sat at a work bench in the shed, next to his ride-on mower, his tractor and his quad bikes. The air sickened with the smell of wisteria blossom, followed by a faint diesel chaser. The casuarinas along his creek are massive and the wind mourns its stories through their needles. As I shaved off two sides of the flowering wisteria cutting and pushed it into the valley groove, taking care to keep it to one side, I tried to leave the chapel windows, just clear of the base plant. This is where you can just see the circular shaved tip of the cutting above the bottom plant line. Then, I learned to wrap the two parts together with binding tape and tie it off. A bandage. Triage for two traumatised plants.

Only about 20% of these grafted plants survive the ordeal, but buyers want a flowering pot plant. They won't buy one without a visible blossom. As I surveyed the five plants that I had masterfully grafted, a wave of nausea overcame me. I was conflicted by what I had done. Interfering with nature – 'improving' and 'augmenting' nature – is a cardinal sin, is it not? The wisterias had no say in their augmentation. It made me think about what fools we humans are: we will only buy a flowering pot plant, not satisfied with the time frame of real plant life. We will only eat big red tomatoes, not happy with the runts of the harvests. The constant human desire for brighter and bigger is the cause of the grafting procedures. The drive to intervene and create interventions is a specifically human trait. It is a form of mastery and control over nature but it is also a desire to re-connect and merge with nature (see chapter on hybrids). We can't help meddling. Perhaps it is a perverse human endeavour to keep proclaiming our hubristic pride by imposing ourselves on other species.

The process of grafting is not limited to the nursery world. A US artist who works with robotany or, as she calls it, dendroids, is Roxy Paine. The work, *Graft* 2008–9, is a stainless steel structure, a meticulous representation of a tree exhibited at the National Gallery of Art, Washington DC. It is a silvery metal tree, full size, and emerges from an area of grass outside the gallery. Its sci-fi reflective surface is hugely desirable, its tendrils turn and twist. It has no leaves. Its uncanny strangeness in such a normative garden site reminds us of the persistent desire among humans to change and recombine part of nature for our betterment, for our aesthetic pleasure.[23] Humans: we can't stop ourselves from participating in and changing nature. Not only do we inhibit natural growth

---

23   Roxy Paine, *Graft*, https://www.nga.gov/collection/paineinfo.shtm, Accessed 6 January 2017.

rates and processes but we also endlessly mimic the natural processes that we work so hard to impede. This is the paradox of human relations with plants.

The Crayweed

There are two artists who have been working to support the rights of plant life beneath the sea, who are drawing our eyes and hearts and minds back to nature. In their work is the suggestion of a contract of awareness, a contract of care. Jennifer Turpin and Michaelie Crawford are Australian artists who collaborate with nature. Their public works are sculptures that function alongside the community in the built environment or at the edges of land and sea. The artists' practice focuses on the movements and lost formations of the natural world. It is the vitality and changeability of the atmosphere and natural elements that drives their work.

Harnessing wind, water, earth and sun, but unleashing it at the same time, Crawford and Turpin use turbines, sundials and waterfalls. These moving and changing parts create a whole structure or series of related parts that highlight the unexpected processes of the natural world. As self-proclaimed collaborators, they are fiercely aware of the problematic human 'collaborations' with nature that have occurred in the past and continue to happen.

Turpin and Crawford are eco-punks in the sense that they are advocates for the regeneration of marine life and they make use of their careers in public art and community projects to spread the word about various crises in the environment. Their point of distinction is their celebratory approach to aesthetics. They are creating artworks in an era where many have lost faith in the arts and when the trust that art can make a 'difference' has become increasingly difficult to prove and harder to find funds for. If eco-punks are those working on the fringes of public and private spheres, of structural and ephemeral change, and of the past and future, then this duo are eco-punks. Their recent 2016 *Sculpture by the Sea* event was punk! Along the coastal walk from Bondi Beach to Bronte Beach, the artists wove fluorescent yellow safety fencing along the barriers and floated a matching yellow buoy off the shore. This installation snakes around the coastal walk, an intrusion in the day-to-day operation. To use a C.S. Lewis expression, the installation 'uglified' the scenery. Plastic, yellow, clunky: a warning to 'stand clear.'

But that wasn't the punk part of the action. The artists rallied hundreds of kids from local schools to make sea outfits – sting rays, octopus, crayweed – so they could promenade and then sing songs on the headland as a choir, as an orchestra, as a strong voice of hope. As the musicians' voices and instruments soared, I heard the sounds as a wailing. A sea calf or a migrating school of whales. I could even imagine the shuddering reverberations underwater as the children's voices travelled out in all directions. A dirge. A lament.

This social praxis project, *Operation Crayweed*, (see figure 7) was a collaboration between Turpin and Crawford and the University of New South Wales Department of Marine Ecology. It was also a community event: a gathering of kids, botanists, engineers, musicians and crayweed experts in action. An engaged and activist move to raise the chorus of young voices, which were carried by the brutal coastal wind across the waves and out to sea. These grassroots events are more powerful than first appears. Let's say there were three hundred people involved in the sourcing of outfits for the kids, helping the making of the outfits, helping with set-up. There was the local mayor, board members of the Sydney Institute of Marine Science who supported the project, and school parents and art lovers. All of these people gathered on the headland. Each one takes home a memory of that day.[24]

FIGURE 7   *Jennifer Turpin and Michaelie Crawford. Operation Crayweed – Art, Work, Site 2016.*
             Performance, plastic buoys, plastic fencing etc.

24   School children from 5 local schools; Clovelly Public, Coogee Public, Rainbow St Public,
        Randwick Girls and Boys High Schools participated in a program of both science and art
        workshops to inform and inspire them about the Operation Crayweed project, extend-
        ing the project's reach into the community. With the marine scientists from USNW and
        SIMS, the school children visited the crayweed test plantation site at Long Bay, then in the

The second reason Turpin and Crawford are eco-punks is because this project resulted in the regeneration of the local sea plant, the crayweed, along the coastline. Once thriving, it had since disappeared. Botanists and marine ecologists worked together to find ways to replant the underwater forests and regenerate the watery weed ecosystems.

From Cronulla, south of Sydney harbour, to Palm Beach, north of Sydney harbour, crayweed (Phyllospora comosa) has been replanted. The sewage facilities along the shoreline caused the damage to the crayweed populations. The development of a test facility for crayweed plantations at Long Bay in southern Sydney is interesting for two reasons. One, Long Bay is known for its human correctional facility. Now, Long Bay is the site of rebuilding ecosystems and regenerating crucial sea life, for the certainty that other elements of sea life can thrive – such as the sting rays, the rock formations and their myriad sea slugs, anemones, etc., all the way up to the shark populations or the migrating whales. Secondly, Long Bay jail is a place of rules and regulations, a place where bad people experience punitive treatment. Those locked up are criminals, from thieves to rapists. We are able to punish those who cause damage to other humans. Can we create a punitive system where those corporations and companies who do not repair the land they use, according to basic guidelines (which are there but rarely enforced), are also incarcerated?

Along with marine ecologists and the community, Turpin and Crawford are redressing the imbalances between humans and nature: quiet contractual activity. By this, I mean that they are not only participating in the research into how the environment was affected for all these years by sewage runoff, etc. They are also participants in redressing or making reparations for that damage. They created a schedule for groundswell action. They punked the boardwalk. There is a reference, in this work, to the original performative environmental artist of the Sydney coastline, Christo, from 1969. This was another art event that pushed the boundaries of legality and environmental aesthetics.

---

laboratory learnt more about the science and issues of marine habitat and biodiversity. With the artists from Turpin + Crawford Studio, they made wearable marine creature artworks and through the creative process they focussed on imagining the life of a marine creature who lives amidst crayweed. These costumes were then paraded at the project launch of the art-work-site at Bondi, after students from Bronte and Bondi Public Schools performed musical numbers including an adaptation of 'Yellow Submarine.' On the last weekend of Sculpture by the Sea, ocean swimmers from 4 Seasons Swim performed a synchronised swim to celebrate the successful planting of the new crayweed. http://turpincrawford.com/node/76 Accessed 6 January 2017.

Christo wrapped the cliff-face of Little Bay Sydney with fabric, alongside his partner Jeanne-Claude. One million square feet of rock was wrapped, using a team of abseiling assistants. The sheer ambition of the wrapped coast was, and still is, an epic story of Australian art history. But, like Turpin and Crawford's crayweed project, it incorporated a massive team of people. It was another community project that brought attention to the environment, to the way the coast receives such a hammering from tides, surf, wind and storms. The use of erosion-control fabric in Christo's work also brought to bear the devastation of damage caused by humans, purposefully and inadvertently. After the massive project was finished, all the materials were removed and recycled.[25]

Misapplications: Morality versus Ethics

These artworks had the capacity to change our perception of the coast, which is the role of the plant contract. Whilst visiting Crawford/Turpin's artwork, I wondered what the agency of the coastal rocks might be. I started to wonder what the distributed ecosystems lying beneath the sea looked like. This is the entelechy of eco-art: the realisation that something – in this case the natural coastline – has outstripped its own potential, that it has achieved a force beyond our expectations. Art, both private and public, socially engaged and distributed, helps make sense of this baffling philosophical question of humanity's place in the world, the natural world. Art often has the capacity to throw moral light on an issue too.

Michael Marder's distinction between ethics and morality play a part here. Marder believes ethics is a way of life, whereas morality is a set of rules and norms for appropriate behaviour.[26] This is a necessary distinction when pondering plant-thinking. This chapter is a nod to the moral rights of plants. It no longer makes sense, given recent plant science discoveries, to relegate plants to a lowly ontological status. Therefore, it is reasonable that the law follows suit ... and legislations and constitutional change. Australia's environmental law policies, like its woeful carbon emission reduction efforts, are on a scale from stagnant to retarded – which is ironic in a country with so much valuable land.

---

25   Christo, *Wrapped Little Bay*, 1969. https://www.artgallery.nsw.gov.au/collection/works/
      152.1971/ Accessed 6 January 2017.
26   Michael Marder, 'Ethics and morality,' LA *Review of Books* web site, http://philosoplant
      .lareviewofbooks.org/?p=177 Accessed 6 January 2017.

It is both the ethics and the morality of plant life that this discourse addresses. Marder's points ring true: attempts to create practical applications of plant thinking 'can quickly and easily degenerate to inane and self-righteous moralising.'[27] So where might that leave the law, or specifically earth jurisprudence? Maybe we can consider Rousseau's 1762 social contract, a text that still has substance today. The community members give up some of their sovereignty in return for care from the state.[28] These days, little has changed. We have to put our bins out on the street on a certain night so that the rubbish will be picked up. We have to accept and pay a speeding fine if we drive too fast. We have to pay taxes so that we can have access to state and federal services. This is so obvious to us now, because it's easy to forget how recently this kind of system developed.

When Rousseau hailed natural landscapes and honoured the workers, whose connections to the land were immediate and tactile, he created a new perception of the role of people in terms of their moral behaviour to one another. His ideas influenced the French revolution, but he is important here because he was aware of the loss of the innocent man and woman – who might not be able to read and write but instead had an intimate, moral and mutual relationship with nature.

The problem is that morality and ethics have developed over time. The ethics of plant-thought, Marder explains, is amoral (not immoral) and is not an application of immutable principles but involve two stakeholders – plants and humans – where both species intersect without rules and norms. Marder bravely states that hollow moral applications such as 'respect for plants' are futile. I believe this is much like the concept of 'hope' or 'positive utopian visions' in climate change issues, which do not achieve results but also prevent action due to a sense that we have relinquished control and only have the abstraction of hope left to us. I am chastened by Marder's warning that such moral codes are no more than 'abstract desiderata.'[29]

If we live ethically, according to Marder's thinking, then we forge a common destiny, whatever that may be. Matthew Hall calls forth the darker side of these debates in his paper on plant autonomy, by reminding us that much of the adverse reactions to plant ethics is that plants are not considered autonomous

---

27   Michael Marder, 'Ethics and morality,' LA Review of Books web site, http://philosoplant
      .lareviewofbooks.org/?p=177 Accessed 6 January 2017.

28   Jean Jacques Rousseau, *The Social Contract,* 1762, https://www.ucc.ie/archive/hdsp/
      Rousseau_contrat-social.pdf Accessed 6 January 2017.

29   Michael Marder, 'Ethics and morality,' LA Review of Books web site, http://philosoplant
      .lareviewofbooks.org/?p=177 Accessed 6 January 2017.

and therefore the conclusion is that any ethical considerations are 'silly.' He also details that applications of ethical treatment of plants might compromise agriculture and even the freedom of plant scientists to kill plants or have their reproductive abilities limited.[30] This, an argument put forth in *Nature* journal, is likely to be the point of view of scientists working in this field.

The point is, of course, that there are several stakeholders who might find the idea of plants having more sensory and cognitive-like capacities than previously realised deeply troublesome. Hall refers to Baumann's proclamation that morality is not about laws and regulations but about group behaviour.[31] Plants' clear and active perception of resources has been established in the plant world, particularly the roots, so Hall finds the non-autonomous argument against plant ethics difficult to swallow. He says,

> By continuing the exclusion of plants from moral consideration, we are intentionally treating plants as less than they are, so that humans may use them without restriction. In human-plant interactions, therefore, moral action may necessitate behavioral changes in the human, which may in some circumstances restrict human action.[32]

If we allocate moral rights to plants, or any other element in nature, there will be stakeholders who will have to work hard to prove there has been no breach in duty of care. Such stakeholders are those who gain from plant harvests in agriculture, even genetically modified farming (think of Monsanto), and those who harvest crops for farm feed and for human food production. It might, theoretically, be in these stakeholders' interests to belittle the philosophy and legality of plant moral rights in that case. In 1972 Christopher Stone asked whether trees should have standing.[33] Children, black people, Jewish people and corporations have rights, Stone argued and that, 'Thus, to say that the environment should have rights is not to say that it should have every right we can imagine, or even the same body of rights as human beings have.'[34] Stone is sensible and unemotional in his invocation of rights for nature. Yet,

---

30   Matthew Hall, 'Plant Autonomy and Human-Plant Ethics' *Environmental Ethics*, Vol 31, Summer 2009: 169–181.

31   Zygmunt Bauman, *Postmodern Ethics*, Oxford: Blackwell Publishing, 1993, 61.

32   Matthew Hall, 'Plant Autonomy and Human-Plant Ethics' *Environmental Ethics*, Vol 31, Summer 2009: 169–181.

33   Christopher Stone, 'Should Trees Have Standing: Towards legal rights for Natural objects,' *South Californian Law Review*, Vol 45, 1972, 450–501.

34   Ibid, 457.

forty-four years later, there has been little legislative change. Marder is correct in his warning that to make legal leaps from morality to legality is problematic. However, there is room to protect certain elements in nature from the adverse behaviour of humans, just as we protect endangered bird and animal species in the same vein.

Rousseau's social contract was based, as mentioned before, on the idea that to have protection from the 'state' as part of a community, we must give up certain individual sovereign rights. For plant studies and for the potential of plant rights, first there needs to be a discourse around why plants are important beyond their production and use. Common law, as Stone explained, denies the rights of natural objects. The merits and value of the natural object are difficult to prove in a court of law. The time, trouble and cost of seeking reparations for damage done to areas of the natural world are a major disincentive. The only way perceptions of nature will change in an applied way is for a case to come to bear in the Australian legal system, for instance and, being propelled onwards, all the way to the High Court. For the 'state' to acknowledge the rights of plants (nature), something must be lost.

In a paper on respect for nature Paul Taylor explains,

> how the taking of a certain ultimate moral attitude toward nature, which I call 'respect for nature,' has a central place in the foundations of a life-centered system of environmental ethics. I hold that a set of moral norms (both standards of character and rules of conduct) governing human treatment of the natural world is a rationally grounded set if and only if, first, commitment to those norms is a practical entailment of adopt-ing the attitude of respect for nature as an ultimate moral attitude, and second, the adopting of that attitude on the part of all rational agents can itself be justified.[35]

The language of *respect* has no force, no impetus, on its own. Even if the terms of that respect are made clear, the clarity of what we mean by respect for nature needs to be excavated. In contrast to human actions affecting nature, Taylor is taking up a position that avoids these issues as human rights or as human values. He rejects the notion of the superiority of humans, preferring a biocentric attitude to nature. As Paul Talyor explains in *Ethics of Respect for Nature*, it is unreasonable to judge nonhumans by using a human concept of civilisation and knowledge as the basis for such judgement.

---

35   Paul W Taylor, 'The Ethics of Respect for Nature,' *Environmental Ethics*, Vol 3, 198.

New Connections: Different Kinds of Eco-Punks

Donna Haraway begins her book *When Species Meet*[36] by taking a walk in a forest where her friend sees a rock shaped as a dog. From this point, she launches her discussion in animal studies and the rights and delights of companion species. What beckons the reader into her discussion is her ability to draw connections between human and non-human beings, both animate and inanimate, biotic and abiotic. Haraway uses personal experience to extend the theory of dog-human relations and how different species are not fixed units but are constantly becoming something else. There is an implicit sense that the haptic connections between human and dog are, in fact, the logic of each other's histories.

In her paper 'Sowing Worlds,' Haraway writes about the great Ursula Le Guin's carrier bag story – *The Author of the Acacia Seeds* 1974 – and her cast of therolinguists; that is, antspeak and rocktalk: a future non-human language. Or is it from the past? Ants and acacias, she tells us, are diverse and well-populated groups and they are also homebodies and world travelers at the same time and have an effect on terraforming. Haraway reminds us that my discussion in this book is no more than a metaphor, or a story. She says,

> Planting seeds requires medium, soil, matter, muter, mother. These words interest me greatly for and in the sf terraforming mode of attention. In the feminist sf mode, matter is never 'mere' medium to the 'informing' seed: rather, mixed in terra's carrier bag, kin and get have [sic] a much richer congress for worldling.[37]

This is the task for the plant contract: to find a means of connecting humans with plants more closely, in order to learn and re-shape our lives. This necessarily requires a change in moral and legal perceptions of the vegetal. Just as Haraway sees the secret beauty of a deep relationship with her dog, we can learn the mysterious value of the plant world as source of inspiration rather than utility, as models of efficiency rather than a wild mass to be tamed. Where Haraway tells us that dogs are not just to think alongside, they are fleshly and material beings, so too is the plant world a biotic and agented world of thriving activity and communication that has been around longer than humankind.

---

36   Donna Haraway, *When Species Meet*, Minneapolis: Minnesotta Press, 2007.
37   Donna, Haraway, 'Sowing Worlds: A Seed Bag for Terraforming with earth Others' in Grebowicz, Margaret and Merrick, Helen. *Beyond the Cyborg: Adventures with Haraway*, New York: Columbia University Press, 2013, 4.

The ethics that emerge from a new perspective on plants is a vast area, where environmental lawyers rub shoulders with indigenous shamans, eco-feminists and academic botanists.

Karen Houle also writes about the ethics of animal, vegetable and mineral or 'the case of becoming-plant.'[38] Adding to this discourse developed by Haraway, Houle and others are using 'becoming' as a means of dealing with new information about other species and therefore throwing into question the relations of 'knowing' that we previously relied upon. Becoming, as pronounced by Deleuze and Guattari, is a way of avoiding imitation and of effacing functionality or productivity which obscure real being. Houle, like Marder, is interested in expression rather than representation.[39] She qualifies that her concept of a plant-becoming is 'to make evident that these vegetal modalities express genuinely different, rather than nifty vegetal-variation on, our dominant modes of enacting communication and our dominant ways of thinking about what communication is and is the in the service of.'[40]

Much of the language in Critical Plant Studies (by Marder or Irigaray) focuses on growth, excrescence, appearance and constant changing as we connect with different species. There is movement, change and growth. Yet it has occurred to me that we may be forgetting an important element in the change of constant growth – that of stopping, of resting, of not growing but, instead, of waiting. Or even ... of failing. Not all plants defy death, not all plants flourish. Deleuze and Guattari explain becoming as non-progression or non-evolution. Instead it is an alliance.[41] But even alliances take work, and develop and change. As Houle asks, are we really truly interested in becoming-plant or in plant-thinking? Or are these argument lines of Critical Plant Studies no more than a different way of placing plants somewhere, somehow – morally, ethically, ontologically. Are we heading towards another kind of failure?

The Aesthetic of Heavy Failure

Our tardy efforts to be gracious about, and grateful to, the natural world for giving humans succor and produce, sits heavily on many human hearts. It's

---

38   Karen Houle, 'Animal Vegetable, Mineral: Ethics as Extension or Becoming: The Case of Becoming Plant,' *Journal for Critical Animal Studies*, Vol. ix, Issue 1–2, 2011.
39   Ibid, 97.
40   Ibid, 98.
41   Ibid, 110.

important to demand a different perception of the natural world, and constantly so. Darwin tried to draw public attention to the complexity and intelligence of plants in his 1880 publication *The Movement of Plants*.[42] He wrote about plants and their seedling movements, their mature movements, their sleep movements. The movement of climbing plants takes his attention too. The most exciting point he makes is that the radicle tips of the plant act like a brain. Now those radicles are radical, because we now know root systems have communication functions far superior to previously believed.[43]

The history of neuroscience's relationship with plant life has been established, via the work of the Enlightenment thinkers.[44] Yet we, in the Western world, have not paid attention to these efforts as keenly as we might. We are blinded by the disjointed difference in time frames. Plants grow slowly, move imperceptibly. The energy and business of their lives are invisible to the human eye that can usually only see potential, profit and individual gain. There are ecology monks, however, in contemporary society who are working against the grasping for social 'improvement' at the expense of nature – they are actively taking a spiritual stance against the de-privileging of the rights of trees. Phrakru Pitak Nanthakthun is a Thai monk who travels to Laos and Sri Lanka to ordain special trees and this role is intended to provide practical and spiritual guidelines for the ethical treatment of trees.[45] We, as humans, desire to 'flourish' but we are caught up in a defensive and destructive means of flourishing that prohibits the flourishing of other species, often. So we build homes and cities, and clear trees for highways to better flourish – without really acknowledging what price is paid, what areas of nature will never be reclaimed, forgetting about the carbon that trees absorb, the clean oxygen they emit.

Grasping and clinging to concepts of continuous economic growth causes a sense of thinking that we never quite get what we need. Instead, to centre ourselves – away from that grasping mentality – and taking care to look at plants in a specific way (with wonder, appreciation and thanks) allows a different relationship. It becomes a relationship of gentle mutuality.

---

42   Charles Darwin, *The Movement of Plants*, London: John Murray, 1880.

43   Ibid.

44   Baylee Brits, 'Brain Trees: Neuroscientific Metaphor and Botanical Thought' *The Covert Plant*, Santa Barbara: Punctum Books, 2017.

45   Susan Darlington, *The Ordination of a Tree: The Buddhist Ecology Movement in Thailand*, Albany: SUNY Press, 2012.

Plant Art Emerges from Human Failure

We have failed. We, the humans, have not been cautious enough – we have not taken care. Our failure is moral. Our failure is critical. Our failure has been an inability to apprehend climate change. But this is not the time for doom and gloom. It is a time to honour the eco-punks working in this lacuna of morality and plant life. This human failure is made clear by eco-punk Australian artist Anna McMahon. She brings attention to past representations of plants and articulates a commentary on related failures (such as the Catholic Church system). She also brings attention to the fragility of plants and the paradoxes of our 'keeping' them. We cut flowers and try to sustain them for as long as possible inside our homes. We forget how strange, even uncanny, this practice is. But is it immoral? In a similar mode to gardening and grafting, landscaping and weeding, the concept of bringing flowers indoors is interesting because it is another means of absorbing the natural into a space of control, order and civilised structure.

McMahon brings cut flowers into the gallery space, but for a different purpose. She makes installations around plants and sport. For McMahon, it is a pursuit of nothingness, an aesthetic of failure. Her work is relevant to this concept of plant ethics or bio-rights because she incorporates plants as metaphors of concerted energy and breath, which are also embroiled in sporting activities. She is fascinated by failure – at sport, at life, at art and at that ludicrous process of living. Plants represent purity, control, emergence, change, decay and strength/fragility.

First, the installations. McMahon plays with the heaviness of failure as a quality in the world. The global malaise that exists for those of us who wish to take better care of the environment – but are embattled by bureaucracy, politics and economic rationalism – is evident in her work. She places basketballs in court hoops, with wattle and red hot poker flowers jammed into the backing boards. There are subtle sexual undertones in the erectness of the plant stems, which slowly become flaccid and wilted. There has been no shot made, the works seem to say, but no shot has been missed either. The action is caught mid-throw.

Other works reveal McMahon's interest in performance, endurance and duration. As the sporting action is caught in the middle or in between, still the plants continue to change. They breathe on until they slowly die and smell of fetid decay. Moisture and energy have been exhausted. One of McMahon's plant artworks comprises a long stemmed kangaroo paw balanced in a tiny finger bowl of water. These bowls are reminiscent of McMahon's past association with the Catholic Church as a child. The bowls were used by the priests to cleanse their fingers before distributing the sacramental bread (the Eucharist), also known as the host. Here the nervously balanced stem (against the end of a wall) could topple at any moment. This brings to mind the fragility of religious logic within the framework of faith. There is not enough water to keep the kangaroo paw alive; the church does not have enough sustenance, the plant will die and fall: it is only a matter of time. This is the narrative of the art work.

In *Untitled #22* from the series *The world is weary of me, and I am weary of it,* McMahon fixes a neon hoop-ring to the wall and hangs an orchid (white Phalaenopsis). This hanging orchid has a vaguely phallic edge to it, yet it is labia-like too. The multi-gendered sexuality of the flower highlights the femininity of the neon ring which could be read as a womb or vaginal orifice. Orchids have female associations, which is interesting as they have a histori-cal significance as being owned by men, propagated by men, painted by men, and kept and exhibited in greenhouses by men. There is strength and fragility to McMahon's work which is impressed on us by her use of the biotic plant matter and its tension with iconic sporting equipment.

By drawing our attention to the dilapidation of the cut flower, its failure to survive out of the ground, McMahon creates an expression of vegetal life that is respectful and illuminates the paradoxes of our strangely unnatural human lives. Is it immoral to bring cut flowers from their place of growth and into the home? Whether the answer is yes or no, it is a metaphor for the more impor-tant work to be done in this broader area of plant and nature rights. Once we can achieve similar changes to legislation, as the Te Urewera Act did, there may be a clearer view of how to proceed with other plant rights issues.

Feminist Plant Empathies

Later in this book, I follow the story of the water lily as a metaphor for differ-ent relations between human and plant. This idea serves as a segue to new perceptions of plant life as models for new perceptions of human life. Catri-ona Sandilands writes about botanical queerness as a way of telling stories.[46] She applies plants' destabilising taxonomical existence in the 18th century to current questions of agency now. These issues present plants as 'queer players.' Plants are not just objects of concern for Sandilands but are subjects of futurity and agency. The art world is the perfect place for queer performativity to be

---

46   Catriona Sandilands, 'Botanical Queerness,' conference talk, https://vimeo.com/ 120086842.

expressed. Radical non-association and the exceptionalism of the exceptions (rather than the hetero-normative) are the qualities of much contemporary art – which is focused on performance, the environment and experiences that illuminate a point of difference.

Eco-punk Australian artist Cat Jones completely dismantles our perceptions of plants in her performance experience. Her concepts of plant life in us, rather than behind us or in front of us, are a queering. Jones disrupts our sensory perceptions and adjusts our sexual self-identities. She does this by transforming our bodies, from man to woman, human to plant. The transgender trans-species model of her work fits within an aesthetic of queer play as well as the associated politics.

Inside the University of South Australia's art-specific building, at the end of a gallery space was a black tent, its entrance curtains swaying gently in the air conditioning. This interactive work was participatory and sensory. In Jones' words, *Somatic Drifts* 'explores trans-human and inter-species empathy. It enables the participant to experience the bodies of other entities, through body illusion and touch. Bodies and narratives travel beyond their own boundaries to ask what realm does the human exist within? How far can we drift? What can this drift enable us to change?' That sounded exciting but was it all smoke and mirrors?

Cat Jones offered me her greeting card as we sat outside her black tent, all part of her consultation process. The card was small and had a botanical illustration in green ink of a half-skeleton, half-tree figure. It was a hybrid human-plant. The tree-half's foliage extended from hip to head, its trunk matched the skeletal leg bones and its root system extended like a human foot. "Smell it," she urged. She had spent time in New York at a perfumery to make a series of plant-related perfumes. The card smelled of the earth: manky, damp and fecund.

Inside the tent was a large, spongy platform bed with black covers. The surface was warm from the downlights. I lay down and stared up at the full body-length screen above me, upon which real-time footage (of me) was projected. It looked no different from a mirror, so did not provoke any anxiety. Ambient sounds drifted through the head phones Cat had given me. She touched my hands and feet, my shoulders and knees, and I could see her disembodied hands (her black clothing made her hands seem independent of her body) moving around me. Slowly, my body was superimposed with another body, the hands were doubled (adding previous footage to the Realtime footage) and I began to apprehend the touch of hands that were recorded. I didn't feel their touch, but I expected to and so there was a ghostlike tingle where I expected to be touched.

Cat used smell and haptic sensibilities to draw me into an empathic sensation, where my body and that of another person (in the screen above me) became one. It was when she introduced the branch of a tree, along the flank of my left arm, that I really began to change my previous understanding of my body. The faint rustling of leaves in my earphones, the residual aroma of plant matter behind my head and the image of Cat running her hands along the branch of a leafy tree, where my arm should have been, caused a fissure in my thinking.

It wasn't that I began to think like a plant. I felt the same – coherent and conscious – but the sensation that I was inside a plant, at a height, was intensely strong. I was ivy, growing high up the trunk of an old oak. As Cat touched my leaves, I knew how a plant felt when a child approached and touched it. I knew how a plant's leaves moved in the breeze, sensorily. I felt like I was encased within the branches of ivy, looking down at the grass scattered with fallen leaves at the base. I was ivy.

Many other participants were equally suggestive during their time in the black tent. There was video footage on screens outside the tent of previous visitors. Some said they felt they were dying or in a state of decay. Others felt alarmed by the sensorial suggestion that they had co-embodied someone of the opposite sex. They 'felt' the man's muscles. *Somatic* refers to bodily being, particular to the energies and activities of the body, but not as separate from the mind. My experience of *Somatic Drifts* was not a duality, a dyad or a dynamic cut between body and mind or between thought and memory. Instead, this experience was a trick – a manipulation of senses to alter my sensibility, to change my understanding of what it is to be a human in the world. It was a radical shift of perceptive experience and anthropocentric expectations.

Val Plumwood sees a connection between the plight of plants and the plight of women, which are inherent in the implications of Jones' *Somatic Drift*. Plants and women have been remaindered in their domestic situations. They are unseen, disregarded or marginalized. Plumwood's eco-feminism refers to the mastery of the human (its desire to master and conquer) nature and the concomitant mastery of male patriarchy (its desire to master and contain) over women.[47] Is it possible to break the habit of perceiving women and plants as a source of consumption? Is it possible to relate to plants and women without the confines of agriculture, without the constructs of gardening and without the subjugation as less relevant or inferior? This tie between plants and women is the eco-feminist structure and it would be remiss not to ask if this is no solution but a convenient metaphor. And would there be further confines and

---

47   Val Plumwood, *Feminism and the Mastery of Nature,* London: Routledge, 1993.

obstructions created by lumping them together in the same pot? Whatever the answer to that subjunctive – what if – the more precise question is whether eco-feminism can shine a light on the (in)ethics of considering plants as onto-logically less relevant. Does the history of feminism, with its hard-won victo-ries and its failures, really contribute to the discourse of the status of plants? Are we talking about apples and oranges, to use a planty pun?

Karen Warren suggested there needed to be an overview of ecofeminist phi-losophies.[48] She connects women and the environment in several ways. The conceptual relations between women and nature being historically conceived as a dualism is part of the issue. Mother Earth, Terra Mater, Mother Nature ... relentlessly referred to via the pronoun 'she.' Nature as a benevolent and fer-tile being; a caregiver, a guardian, a custodian. These are different words from those more often associated with the masculine – guard, gatekeeper, ruler, sov-ereign, provider.

Plant Ethics is Silly

For ethics to be implemented, the collective wish must be there first of all, and then the communal will to constitute change. Matthew Hall draws a con-nection between the lack of autonomy afforded the plant and the subsequent implication that plants having ethical rights, as ethical beings, is impossible.[49] He rages against an article in *Nature*, where plant ethics are cast as 'silly.' Hall does not call for these concepts of plant ethics to be legitimised in legislation, nor that they need to be codified as law.[50] The 'good' of the individual nonhu-man organism is an imperative of morality. Striving for the best, for the survival of the one and of the 'all' is all that matters. The 'good' of the plant suggests not only an individual ethical being but, to me, seems to echo Rousseau's concept of the collective good in his social contract.

Nealon's book *Plant Theory* energetically traces Derrida's ethical inves-tigation into animal studies and sets those writings within a wider ethics of nature. Nealon refers to Derrida's distinction between living and surviving. Mere survival does not necessarily qualify as 'life.' They are alive, Derrida, says,

---

48   Karen Warren (ed), *Ecological Feminist Philosophies*, Bloomington: Indiana University Press, 1996.

49   Matthew Hall, 'Plant Autonomy and Human-Plant Ethics,' *Environmental Ethics*, Vol 31, Summer 2009.

50   Ibid, 170.

in a biological way but not in the sense that they have a sentient life. There is the power of growth, the possibility and potentiality of life in plants.[51] Marder refers to this growth in his book when he mentions the extrusion of plant life – the disgusting and inappropriate growth on the limb of a tree, on the branch of a plant.

There is no vegetal desire for a higher state or higher knowledge in Derrida's *Glas* for instance but is this an appropriate set of criteria for deciding its worth. Plants' only desire is to grow and grow until they grow no more. Is this less relevant than a human who desires sex and love and family until death?

The Eco-Punk Outlaws

Moving beyond the disdain for plant life relies upon a dismantling of seeing plant rights as 'silly.' A plant excrescence is an ugly growth or a monstrous knob. Marder's adoption of the word is a reclamation of appropriate language for plant life. He says, 'An excrescence of plant-thinking, it nonetheless risks turning into a cancerous growth, suffocating the very entity from which it draws its vitality.'[52] This is a timely moment to reflect on the irony of an ethical discussion on behalf of a species that may or may not be interested or able to benefit from such a discussion.

Even if plants do not require our human legal services, it still makes sense to protect all of that which aids us. As mentioned earlier, in 2008 the Swiss government made an amendment to acknowledge the moral rights of plants. Now, Switzerland's law takes a biocentric position endorsing 'the belief that all forms of life are equally valuable and that humanity is not the center of existence.'[53] Ecuador also amended its constitution, ratified by referendum, giving recourse to its rivers and forests in their courts of law. As of 2008, their new amendment now includes 'Rights for Nature,' giving legal authority to those who defend the vital cycles of ecosystems. The ecosystem can be named the defendant.[54]

51	Jeffrey Nealon, *Plant Theory: Biopower and Vegetable Life,* Stanford: Stanford University Press, 2016, 60.

52	Michael Marder, *Plant Thinking: A Philosophy of Vegetal Life*, New York: Columbia, 2013, 177.

53	Heather Ring 'The Dignity of Plants' *Archinect* 23 March 2009 http://archinect.com/features/article/86646/the-dignity-of-plants Accessed 20 July 2015.

54	'Ecuador Adopts Rights of Nature in its Constitution' *Global Alliance for the Rights of Nature* http://therightsofnature.org/ecuador-rights/ Accessed 6 January 2017.

These provisions are legal endorsements of a changing philosophy of plants: the plant contract. There is no evidence of neurons in plants. However, plant signaling and communication science has been developing in the last decade. The effects on our cognition of plants, the idea of plant cognition and consequently our human relationship with plants has changed and these discoveries are already effecting change in legislative policies. Plants are the outlaws.

Derrida speaks of the sovereign reign of physis (growth) and Nealon[55] believes this is a reference not to the human or the animal but to the plant growth or becoming that is indifferent to man. This is a crucial point that many plant theorists have not attended to yet. Post-humanism works towards a dismantling of centralized human-ness, a collapse of human exceptionalism. Part of this thinking demands that we face facts that plants exist with or without human witness. Yet still we argue back and forth about whether plants are alive, whether plants are sentient, whether plants have a will or agency or a desire for higher thought. What we fail to embrace, over and over, is that plants may not care. Humans still cannot grasp their own irrelevance. Plant relevance, to summarise, lies in moving beyond a philosophical redressing: it is also an invitation to take steps for self-preservation. These two entities, active plant rights and human survival, end up being mutually exclusive.

Crimes against Nature

To elaborate on active plant rights, in 2015 a group of artists collaborated to bring to trial those who had transgressed the rights of the natural world, as well as the extermination of animals and the genocide of humans. ZKM is a cultural institution in Karslruhe, Germany. Their *Globale* tribunal, held in June 2015, brought speakers and thinkers to their foyer where they took their narrative cue from Franz Kafka's 1912 novel *The Trial*.[56] Like the Nuremburg trials, the conference explored methods of holding humans accountable for their actions against the world. There was a film, an installation and a critical trial. This public trial was framed against the exploitation of the earth and associated violence and genocide during the 20th century. The logic followed that during the 20th century modernity had committed violence and genocide. Part of this tribunal included an investigation of the massive annihilation of human

---

55   Jeffrey Nealon, *Plant Theory: Biopower and Vegetable Life,* Stanford: Stanford University Press, 2016, 73–74.

56   http://zkm.de/en/event/2015/06/globale-tribunal-a-trial-against-the-transgressions-of -the-20th-century Accessed 6 January 2017.

life, exploitation of the planet, and extermination of animals and other living organisms in the twentieth century.

This critical prosecution of a period of time, the 20th century, marks an interesting change regarding how to establish guilt and how to achieve reparations for the natural world. Phyto-mining is a growing area of knowledge and industry action. The planting and harvesting of particular crops and forests in order to extract minerals (attaching to their roots) from the earth creates complex ethical scenarios. If this has an easier footprint on the environment, is it a legitimate solution? Or is it a continuation of a hierarchical system of being, where plants are not given equivalent status?

Wild Law

Enacting trials against those who perpetrate crimes against the planet might seem like child's play, but with no other legal recourse the ZKM trial represents the voice of the voiceless, a desire to draw attention to the distant shrill scream of the end. When Rousseau called for social justice, he said, 'The strongest is never strong enough to be always the master unless he transforms strength into right and obedience into duty.'[57] This quote comes immediately after Rousseau's explanation that the first principles are the family unit. Once children are independent of their father, the contract with the family is voluntary rather than natural. Before children separate from their family, the contract is natural.

Rousseau's points are interesting. First, he sees the bond between a family nucleus as natural. Secondly, he sees that strength must be coupled with morality (rightness) and that obedience to one's father can develop into duty once that power shift has occurred. Is this, perhaps, one of the reasons the plant world has been overlooked? Humans are consumed with the processes and relations of power – within the family, within the community, within the civil city and continuing onwards to the power of the nation. The plant world may operate in a similar manner, in terms of the shared information between forest trees.[58] Rousseau says that men can't create new forces but that they can bring together those that already exist and steer them.[59]

---

57   Jean Jacques Rousseau, *The Social Contract,* 1762, 2, https://www.ucc.ie/archive/hdsp/ Rousseau_contrat-social.pdf Accessed 6 January 2017.

58   Suzanne Simard, 'Leaf Litter, Expert Q and A,' *Biohabitats,* Vol 15, Edition 4, 2016, http://www .biohabitats.com/newsletters/fungi/expert-qa-suzanne-simard/ Accessed 6 January 2017.

59   Jean Jacques Rousseau, *The Social Contract,* 1762, 6, https://www.ucc.ie/archive/hdsp/ Rousseau_contrat-social.pdf Accessed 6 January 2017.

We only now know that plants share information that assists the growth of all trees in a given forest. Canopies form to reach the sun and under-storeys manage to grow within that ecological structure, even vines and ground cover. The symbiotic relations between elements of a forest or rainforest illustrate a cohesive and coherent system under the terms of a social contract. These relations are the workings of the family unit where obedience and moral strength (the combination of the morality and strength) are necessary for survival.[60] Rousseau's social contract can be seen, working, in the natural world. Species 'duty' and 'mutual care' is evident in the operations of plant communities. What is less evident in the plant world is the presence of a sovereign power. What might this absence of sovereignty in the vegetal world tell us, in terms of how to conduct our human lives? Could we perhaps be mistaken about how to achieve good governance in our communities? Is it possible that a plant-thinking mode of governing might be better?

In his manifesto for earth justice, Cormac Cullinan calls for the same rights of law for animals and plant life as for human beings. His writing rethinks earth governance. It is a means of focusing on ways of living where humans try to find out what more the earth requires from them, rather than developing technologies that increase the impact of our requirements of the natural world.[61] Cullinan refers to the Tukano Indians of the Colombian Amazon as examples of a people who commune with the dead and other species, and who control population by abstinence and oral contraceptive made from plants. Controlling harvesting, allowing for replenishment, curtailing population growth! These are the very issues the Western world is grappling with, and failing.

In keeping with critical plant studies ideology, Cullinan argues for a 'new understanding of ourselves as part of, not detached from, Earth.'[62] Earth governance or earth jurisprudence varies from the kinds of environmental laws that seek to curtail the negative impact of human activity on the environment. His wild law is an effort to shift the goal and purpose of the governance system altogether. Plants and animals are currently considered property of a human or a 'juristic person' such as a company and therefore are not holders of rights. Exterminating nature is not illegal.

---

60   Jean Jacques Rousseau, *The Social Contract*, 1762, 2, https://www.ucc.ie/archive/hdsp/
      Rousseau_contrat-social.pdf Accessed 6 January 2017.

61   Cormac Cullinan, *Wild Law: A Manifesto for Earth Justice*, Devon: Green Books, 2011, 91.

62   Ibid, 90.

The Natural Solution

Michel Serres elaborated his 'natural contract' and this has been a useful model for analysing artworks, and collaborating with artists in their plant endeavours. We are no more than renters on the planet and we must therefore be aware there is a bond to pay if we leave a mess at the end of the lease. Our worries these days, according to Serres, are weather patterns.[63] Has time changed our relationship to the weather in the short-term rather than long-term? How does that affect 'the long now' – decision-making with and for longevity? *Temps* means both 'time' and 'weather' in French. We still live in time but we no longer live in the weather. Rain or sushine has little impact on how we go about our lives, day to day.

Our collective sense of a future is different today than it was for our ancestors. The way we read the weather now is less related to harvest and fishing but more to do with philosophical incursions and an apparently growing number of natural disasters. Our understanding of weather has become abstracted and disconnected, yet is also aggravated by politicised and histrionic undercurrents. The catastrophic implications of what humans have done to the natural world defy reason. Political unwillingness to make changes – preferring electoral victories and national account balances that are not red and decidedly not green – are affecting the way we no longer understand nature.

---

63   Michel Serres, *The Natural Contract*, Ann Arbor: University of Michigan Press, 1995, 27.

CHAPTER 5

Eco-Feminism: Plants as Becoming-Woman

Much of the writing of this book has been a concerted effort to discuss artworks that speak with the major issues of critical plant studies and as a means of reintroducing ourselves (as humans) to nature. Wastelands, hybrids, robotany, ethics, rights and moral contracts, have monopolised the conversation to this point. But there is another element of this discourse which fires the hearts of many plant lovers and will be developed in this chapter – eco-feminism, a mainstay of the plant contract. My interests are led by Irigaray's thesis that a separation of the human from nature is a kind of non-thinking and causes a lack of living energy. By focusing on a return to the elements of nature – in this case the myriad possibilities that plant life can show us – humans have a means of returning to careful thought. Irigaray says 'from being alone in nature to being two in love.'[1]

The female, in an emergent state of radical reinterpretation of vegetal life, sits at the nexus of art/plant research and a *plant contract*. Nature has conventionally been cast as a womanly figure. Nature is a mother, a fecund vessel, within which life (that is, the human) can grow. This casting of nature as female, however, can be seen as delimiting and may not serve the purposes of becoming woman, of fulfilling both an independent and a collective state of being. That is, constant growth. This chapter acknowledges the history of feminising the planet but reanimates and recasts that process by introducing the water lily as a breathing becoming-woman. The flower, especially the water lily, is an eco-feminist object and has been a motif in art. This chapter refers to French modernist artist Claude Monet's water lily and also the olfactory experience of artist Cat Jones' 2017 *Scent* performance (See figure 8).

An addition to the historical mythologising of the natural world as a woman, there is the chronicle of botanical study in the 19th century. There was an efflorescence of female interest in botanical studies and its scholarship in Britain at the time. Women, for this short period of British history (mostly under strict controls and without the possibility of publication of their own work), were allowed to investigate botany from the safety of their home environments. However, it soon became clear that the sexual information that women

---

1   Luce Irigiary and Michael Marder, *Through Vegetal Being.* New York: Columbia University Press 2016, 84.

FIGURE 8   *Cat Jones. Scent of Sydney 2017. Performance, sound, biotic matter. Carriageworks,*
             Sydney Festival 2017.

accrued via this plant science knowledge was not deemed beneficial for them
and limitations of access were widely imposed by the end of the 19th century
in the interests of social morality.[2]

Botany has been a domain of thought and scholarly engagement that has a
marginal female heritage. These restrictions and constraints are still reflected
in the male-dominated publication numbers, papers written by philosophers
and scientists, which are now in the canon of history. These unequal representations cut across philosophy as well as botany. In one of his books, Michael
Marder writes essays on philosophers who have written on plants and trees
such as Aristotle's wheat, Leibniz's blades of grass and Derrida's sunflowers.[3]
Marder follows ten male philosophers, but only refers to one female: Luce
Irigaray.

Since that publication of plant philosophers, Marder and Irigaray have
worked together on different texts, both scholarly and journalistic, focussing

2   Ann Shtier, 'Botany in the Breakfast Room: Women and Early Nineteenth Century British
    Plant Study' in Pnina Abir-Am and Dorinda Outram (eds) *Intimate Lives: Women in Science
    1789–1979*, London: Rutgers University Press, 1987.
3   Michael Marder, *The Philosopher's Plant,* New York: Columbia University Press, 2014.

on plants or vegetal being.[4] Irigaray always had a deep connection with plants. She has said that thought needs 'to be ready to listen to nature, to the sensible.'[5] By this she refers to the importance of being aware of physical surroundings and material differences between elements in nature. Irigaray draws out the 'nothing of nature in growing' as a taking on of the void which refers to the life of the vegetal world beyond utility and beyond a human understanding of 'nothing.' She warns her readers to distance themselves from experiences that are only immediate, from not always mastering the real. Rather than seeing plant life as objects, only there for our subjective experience, we can think of different time scales and the possibility of different non-immediate experience. By breathing, she suggests, we can continuously re-open the possibility of new growth for life, for desire, for culture and for love.[6] There is a requirement for breath, Irigaray's philosophical legacy, even in this writing of the water lily as a becoming-woman. Plants breathe out and in, contrary to humans' in and out. The synergy is something that humans yield rather than vice versa. Humans gain from plant life, yet plant life yields little from the human.

Sexuality versus the Sexuate being

In terms of sex, plants often have a male and a female part in the one plant. Some mosses have male plants and female plants working together. Some conifers have two types of cones; one is the stamen cone, the second catches the pollen if the wind is howling right. Certain flowers have both stigma and stamen in the one plant. These last self-fulfilling processes of reproduction have enormous significance for a culture where gender and sexual politics are a constant source of queering change and fluidity. A greater awareness of plant difference creates potential provocations for a re-thinking of the world – where individual species have hybrid, undetermined or multiple sexualities.

This text addresses plants and art as provocateurs in a discourse, and in this chapter focuses on the water lily as a plant with aggregated representations, histories, significance and feminist properties. The water lily is a metaphor for woman. It is an allegorical device for drawing attention to the vegetal world. Women and the vegetal world have much in common. The water lily is also a

---

4   Luce Irigaray and Michael Marder, *Through Vegetal Being*, Columbia University Press, New York, 2016.

5   Luce Irigaray, *I Love to You* New York: Routledge 1996, 139.

6   Luce Irigaray and Michael Marder, *Through Vegetal Being*, Columbia University Press, New York, 2016, 96–97.

sexuate being. In this text, I refer to sexuate as the way plants are situated and how they independently exist in terms of sexuality, without being reduced to their sexual processes. Luce Irigaray believes 'cultivation of our sexuate surges is crucial for our becoming able to behave as a living being among other different living beings without domination or subjection.'[7] This is relevant to the water lily and artistic representations that relate to sexuate being and plants. Being relieved of domination allows writers to consider the water lily and its effects on humans as something that can constantly close and open, much like Irigaray's notion of breath.[8]

Approaching this subject matter within the context of the feminine is specific. Humanity's place in the world is centralised. This anthropocentric 'bind' is why, according to Claire Colebrook, 'becoming-woman' is still required.[9] This could be a human-plant alternative, a 'shaking of the tree' to create new political and cultural units of thought. Or it could be a quiet listening for the breath of plants. This is not a frequency that humans can hear but it is a process by which we can become more aware, more respectful.

Luce Irigaray's choice of plant, as discussed by Marder, is the water lily. The water lily has roots and it also floats across the water, a fluidly female surface. It reproduces asexually with the help of insects carrying seed from the anther to the stigma. The water lily cannot be tied down and requires the help of an entire ecological community to thrive. This is an important model for living or for fully expressing life. Breathing through the skin with the whole body, as Buddha contemplates the lily,[10] can be aligned to the way a plant's nodes open and close with each breath of wind. This allows the thing that we are attentive to, to become attentive of us.[11]

Water lilies flourish in clusters and chemically communicate to one another within their community.[12] They can self-reproduce and move across watery surfaces, being both rooted in the earth and waterborne. I am proposing here that if there is a capacity to find metaphorical connections between human culture and natural life, then this allows a space to introduce the water lily as a feminist plant. In a social and political context, the water lily relies on no

---

7   Luce Irigaray and Michael Marder, *Through Vegetal Being*, Columbia University Press, New York, 2016, 87.

8   Luce Irigaray and Michael Marder, *Through Vegetal Being*, Columbia University Press, New York, 2016, 21.

9   Claire Colebrook, *Sex After Life*, Open Humanities Press, 2013, 158.

10   Michael Marder, *The Philosopher's Plant,* New York: Columbia 2014, 224.

11   Ibid, 214.

12   Monica Gagliano, 'The Mind of Plants: Thinking the Unthinkable.' Communicative and Integrative Biology, 2017. http://www.tandfonline.com/doi/full/10.1080/19420889.2017.128 8333.

individual authority, nor is it dictated to by its male counterparts. As with most plants, it has been marginalised and treated as unequal or ontologically less relevant in the past. Vilified for being too passive and too silent, disparaged for having inferior abilities and condemned for being immobile, the water lily nevertheless defies these denigrations.[13] The water lily is a feminist plant.

Luce Irigaray believes we have moved away from the vegetal world and neglected what being alive presupposes. She suggests that it has taken our planet being threatened to finally realise what the basic conditions of living mean.[14] If speaking can't ever be neutral, should we not speak? In terms of the water lily as a feminist plant, it is time to speak. Speak via eco-transmissions. Speak via chemical releases. Speak via the movement of flowers across the water. Speak, irrespective of whether there is a human there to listen. So, the plant contract speaks.

All plants create oxygen through photosynthesis and they absorb carbon dioxide. All animals breathe oxygen and exhale carbon dioxide. Most of us remember learning this basic fact in primary school science. Yet still we have forgotten these crucial points. Plants and humans breathe together. When did we forget this? Irigaray warns of trying to make animals our partners in the universe or, worse still, a kind of human-like version.[15] Humans can't write of things outside our comprehension. However, we can write a response to the independent agency of the water lily.

The Whole Water Lily

The water lily ecology is both an element in nature and an entire environmental structure in itself. Gaia is a concept developed by James Lovelock to engender a form of the living earth, a form that might be seen from space.[16] This concept comprises an earthly 'it' that is alive and moving, constantly decaying and dismantling. The earth's self-making and its limits are intrinsic to the Gaian hypothesis. Donna Haraway acknowledges the mythological elaboration of Gaia

13 Michael Marder, *Plant Thinking: A Philosophy of Vegetal Life*. New York: Columbia, 2013.

14 Luce Irigaray and Michael Marder, *Through Vegetal Being*, Columbia University Press, New York, 2016.

15 Luce Irigaray, 'Animal Compassion' in M. Collarco and P. Atterton (eds) *Animal Philosophy: Essential Readings in Continental Thought*, Continuum, London 2004, pp. 195–201.

16 James Lovelock, 'Atmospheric homeostasis by and for the biosphere: the Gaia Hypothesis,' *Tellus* XXVI 1974.

and the Anthropocene but she also articulates its limitations.[17] These limits might include the descriptions of the earth as a she, as Earth Mother, as Terra Mater. I prefer 'it' or the genderless pronoun 'hen.' The earth is a gender-neutral being that is constantly changing, a system of being that is in a never-ending flux of experience. Symbiosis, developmental phenomena and change through time, Haraway says, were not dealt with in biology in the early twentieth century. Economy and ecology and world-being in the third age of carbon require a different view of Gaia.[18] Gaia is not a 'she,' then, but a 'we,' or an everything.

Referring these Gaian and anthropocentric ideas back to the water lily, the presentation of the lily as a feminist plant may indeed fall into this trap of anthropomorphism – the binding process of only seeing and understanding the world through human eyes – but it is intended as an allegory. Can allegorical or metaphorical thinking be seen as limiting, and does it have the capacity to reduce the importance of learning from plant life? There is a force in 'story' and in 'metaphor' that avoids this danger and carries the relevance of vegetal life, just as the wind carries seeds across the globe, propagating and disseminating dangerous plant ideas.

The water lily is the plant of choice, here, as a way to discuss vegetal relevance to contemporary cultural thought, because it has an idealised history as the Nymphea bride or 'veiled one.' Its long history of representation has created its iconic status and its associated meaning of purity. The cycle of life and enlightenment still reverberate in illustrations, engravings, stories and architecture. It is the divine, the female and the otherworldly.[19] Rather than becoming frustrated by the gendered issues of the bride, Michael Marder, in his reading of Luce Irigaray's writing on the water lily, harnesses and expands them: 'Through attentive fidelity, we become the betrothed of what or whom we follow.'[20] This is a re-appraisal of the history of representation, beyond a mere attempt at de-substantiation. We cannot redress the way the water lily has been identified in history but we can lengthen that discussion into the long now, meaning a view of the future that has longevity. Can we all become collective brides of the material earth? Can we show allegiance to the dirt and stones, the rivers and forests? Can we commit an honourable agreement, a pact of loyalty? Irigaray suggests that consciousness and thoughtful attention, free from cumbersome 'understanding' is exactly this form of fidelity. It follows a

---

17   Donna Haraway, Anthropocene, Capitalocene, Plantationocene, Chthulucene: Making Kin, *Environmental Humanities*, Vol 6 2015: 159–165.

18   Ibid.

19   Michael Marder, *The Philosopher's Plant,* New York: Columbia 2014, 216.

20   Ibid, 224.

deliberate alliance with Aristotlian ideas of growth. The space to be nourished and to grow is implicit to a feminist modality of being.[21]

Despite the conceivable beauty of the water lily as symbol of female deity, other worldliness and fertility, this concept falls short. So too, according to Marder, has our history of humanity: 'In what amounts to a self-betrayal, and a betrayal of one another, we have historically opted for a cunning synecdoche, according to which one half of humanity – men – is interchangeable with human, as such.' Speech, like the 'mutilated-beyond-recognition' state of nature has wandered away from being.'[22] Where Marder refers to men's breathless aspiration to get to the top of the tree, he speaks of an inability to become. The water lily – despite its history of otherness, femaleness and as divine nymph or goddess – reminds us that breath is a respiration, not to be confused with mere aspiration. The motion moves in and out.

Representation and Expression

While carrying this 'otherworldly' divine significance, it is interesting to remember that the iconic images of the water lily in art, culture and design are flat images. Think of religious icons and Buddha contemplating the water lily. So on the one hand, the waterlily is 'idealised' and revered for its iconic value, and yet, on the other hand, its true earthly essence and sexual versatility are kept 'flat' – immobile and static like a picture.[23] Culturally, women have experienced the same. Women have been 'idealised' and revered as icons, yes, but their feminine essence and sexual versatility have been also kept 'flat' within this cultural/social trap, like the water lily.

Irigaray defies the history of divination, marginalisation and fetishisation by breathing the water lily: writing it, thinking it and thereby giving plants, women and earth-care the room to grow and become. This story, woven by Irigaray and Marder, requires a new construction of the character of the lily within a *mise en scène* – where its activities and behaviours can be understood by humans as a condition of living that can be admired and even emulated, not as object but as fellow participant.

The capabilities and capacities of water lilies, aside from representation, are further evidence that plants require critical and cultural examination as

---

21   Michael Marder, *The Philosopher's Plant,* New York: Columbia 2014, 219.

22   Ibid, 225.

23   Gibson, P. and Gagliano, M. 'The Feminist Plant: Changing Relations with the Water Lily.' *Ethics and the Environment,* Vol 22, Issue 2, 2017.

companion species, and that plants require an accompanying shift in human perception of their vegetal status. By addressing the feminist nature of the water lily, I hope to develop a connection between plant biology and art for a better understanding of plants, via a feminist perspective.

This leads us to the community-focused example of an art-plant hybrid that many humans around the world have encountered – that is, Claude Monet's series of 250 paintings, *Nympheas*. Bioart scholar Monika Bakke says 'art can explore the post natural condition that is typical of modern human-plant relationships.'[24] While the *Nympheas* paintings are not bioart (contemporary art using biological matter, process and methodology) they are nevertheless precursors of the bioart genre in their microscopic detail and alchemical painterly process.

At the original Musee de L'Orangerie gallery in Paris, Monet's *Nympheas* were installed in a lower level oval-shaped gallery space. These paintings of water lilies circled around the viewer, creating a privileged community among those art visitors who saw it, and who shared a love for its beauty at that venue and at all subsequent exhibition spaces. Perhaps it was the way Monet created depth and shallowness at once and how this engendered a sensory affect not experienced before. Or was it the movement of colours in kaleidoscopic propositions that continues to work on human neurobiology in chemical ways? There is a strange and forceful power in these paintings of seemingly prosaic subject matter – a force of energy, a tension between the naturalistic details of the natural plant specimens and the matter of the painted surfaces. The lilies sit on the top of the painting ground, floating over the gesso-primed canvas on board – part of the overall picture but also spontaneously independent of it too, a sensation of natural complication. Deleuze says 'sensation is vibration' and the subject in process of becoming is facilitated by the vibrations in the natural world.[25]

The *Nympheas* are known to have a sensory effect on viewers, both as individuals and as a group. Spurred on by communication of their experiences to one another, the art world's love for the *Nympheas* spread since the painting of them in the modernist era. This fame resulted in prints and posters being hung up in every second (or so it seems) doctor's surgery across the developed world. An epidemic of water lilies. Emotions can vary, but it is not unusual to see a fellow viewer become teary or even choked up with emotion by the lilies. This is the spirit or force of the vegetal artwork; the flat surface, made endlessly

24   Randy Laist (ed), *Plants and Literature: Essays in Critical Plant Studies,* New York: Rodopi
        2013, 181.
25   Elizabeth Grosz, *Chaos, Territory, Art,* New York: Columbia University Press, 2008, 84–86.

ecological via layers of applied paint, reminds the viewer of nature but from a distance.

This is an important point when considering what art can do to change our values about nature. How can art show humanity the harm we have done to the planet, the contract we have broken and whether there are any possibilities for remediation of the agreement? Art is more than a representation, more than a means to uplift (although it can do that too). When art is good, it instills a commemorating force. When we view the *Nympheas* today, we see the reminder of what once was. The legacy. The force of this memory is as strong as the cleverly manipulated forces of painting technique that Monet used – colour as mood, layers of paint as an illusory device, and the haze of light that reveals and conceals in a soothing yet antagonistic way. These are the forces exerted by the artist in his studio. They are used not to disrupt but to bind, like an evangelical aesthetician calling together a community of lookers, an aggregate of people who are bound by a connection to the earth's produce and bounty.

The water lily's roots nestle in the silt below the water but its leaves and flowers are mobile, floating across the reflective water surface. The conventional criteria for considering plants as less relevant than the human species, since the era of 18th century nature classifier Karl Linnaeus, are based on vegetal lack of mobility and sentience.[26] It is possible to undermine this latter convention, based on recent discoveries in plant science that support the concept of plant intelligence, cognitive sensing and learning.[27] Mobility and sentience are scrutinised as legitimate criteria for relevance.

Water lilies have capacities and qualities that support a feminist life due to their adaptive option to self-reproduce, and provide an ideal model for inter-species, non-hierarchical ways of living. We can learn models of appropriate ethical and political behaviour from the vegetal world by raising the status of plants from sub-species to co-species. This suggests a model of thought where vegetal life is equal to human life and where the 'feed' or ecology of dependence is a human reliance on plants for survival – rather than mistakenly perceiving the reverse. This is, in part, a reference to the production and harvesting of plant life, its manufacture and economy. Humans perceive the relationship between humankind and crop plants as a process of dutiful and/or technological care (processes such as water reticulation, bug spraying and artificial nutrients) and that these crops survive only if supported by constant human attention.

---

26   Michael Marder, *The Philosopher's Plant,* New York: Columbia 2014, 63.

27   Monica Gagliano, 'In a green frame of mind: perspectives on the behavioural ecology and cognitive nature of plants.' *AOB* 7, 2015.

The current state of agricultural affairs makes this true in the sense that our practices have selectively reduced both the genetic and phenotypic variability of those plant species we grow as crops. By constraining them into obligate annuals designed for uninhibited sex and early death, the process of converting wild species into tamed plants fit for human consumption has enfeebled them, stripping them of their ability to communicate effectively to protect themselves from pests and diseases.[28]

As Gagliano writes:

> In fact as a result of breeding for increased growth or yield in modern agricultural systems, many modern crop cultivars such as cacao, corn and cranberry, to name a few, have lost their ability to produce adequate quantities of volatile organic compounds to maintain the integrity of their precious relationships with the arthropod community, which includes beneficial insects pollinating them and controlling their pests.[29]

Gagliano explains that plants are no longer able to 'cry for help' or emit key chemical information when they are under attack or are suffering root damage, which means they become susceptible. This then continues into a decline, where yearly yield loads are reduced by the billions.[30] That being said, it can also be argued that this perception is skewed and we have forgotten, due to our cultivated urban relationships with plants, that the natural processes of growth occur with or without human contact. The longevity of ancient clonal plants such as the water lilies, whose cells and tissues can survive impeccably for millennia, is indeed an excellent example to illustrate this point. Thus it can be argued that plant life, without human intervention, provides valuable lessons of adaptation and sustainability for co-species living.

We can develop human strategies for how to focus on the processes of being and becoming by observing how plants are and how they change. This establishes a way to understand that being rooted in the earth does not necessarily correspond to being immobile or lacking in the associated processes of reason or decision-making. It discusses how to participate in a linguistic exchange that exists outside human meaning and it seeks to find a way of being that

---

28   Gibson, P. and Gagliano, M. 'The Feminist Plant: Changing Relations with the Water Lily.' *Ethics and the Environment*, Vol 22, Issue 2, 2017.

29   N. Dudareva, A. Klempien, J.K. Muhlemann, I. Kaplan, 'Biosynthesis, function and metabolic engineering of plant volatile organic compounds' *New Phytologist* 198, 2013: 16–32.

30   Gibson, P. and Gagliano, M. 'The Feminist Plant: Changing Relations with the Water Lily.' *Ethics and the Environment*, Vol 22, Issue 2, 2017.

is not hierarchical and not gender specific, but is instead a return to an inner earth, as opposed to the de-privileging notions of 'Mother Earth.'

We see plant or vegetal writing to be a form of feminist writing, a means to develop the idea of the plant world outside of heteronormative values and outside a generalist community. Humans have ignored the hardships of plant life – just as we have ignored the plight of women in terms of equal pay, equal work opportunities, the right to work outside the domestic home and shared parenting duties. These are socially perceptive conditions that women and plants have in common. That is, vegetal writing sits alongside the growth of plants and acknowledges the wealth of knowledge and capabilities of plant life – it does not operate within an authoritative hierarchy, it appreciates active co-species qualities and it understands that all species thrive when they function as communities.

The Woman Moves

Gagliano says:

> In his world-wide famous opera *Rigoletto*, Giuseppe Verdi entrusts one of his most remarkable arias, "La donna e' mobile," to one of his most despicable characters, the Duke of Mantua. While he sings what becomes his signature tune about the woman being fickle, it is ironically, he blissfully unaware who personifies the volatile and fickle character in the opera. With a surprisingly feminist element, the catchy aria illuminates the arrogant and condescending attitude towards a feminine, whose apparent lack in groundedness ('qual piuma al vento'/ 'like a feather in the wind') and inwardness ('Muta d'accento e di pensier'/ 'She changes her voice and her mind' or alternatively translate as 'She does not speak nor think'), relegate it to a place of inferiority and even unconsciousness. By offering a comic moment in the story with the Duke boasting about his superiority to women in the area of stability, could the aria disclose the idea that the declared feminine fickleness and lack of awareness is a projection of a masculine inner instability? If this was the case, we may expect the unstable masculine character of our culture to unnaturally consign its feminine counterpart to a ground of limited or no movement – if successful, the woman is mobile no more.
>
> Women have suffered from conventions of immobility. Whether bound by domesticity, motherhood or the workplace glass ceiling, these experiences have been adequately documented already. What is less known is

how plants too have experienced a de-privileging associated with an apparent lack of movement. The perception of plants as immobile objects in space and time emerges from the superficial and impatient glances of humans and it is – in Francis Hallé's words (personal communication) – "an extremely deluded impression on the part of us animals." Indeed, this perception has very little to do with reality in general, where "nothing happens until something moves" – as pointed out by Einstein. And in actuality, something is always happening in a constant flow of beginning and becoming that makes life. If plants were truly immobile as prescribed by conventional wisdom, then plant life would not be *happening* – there would be no beginnings and certainly, no becomings. And yet, the growth of plants through space and their rhythmic changes across time stand as an unquestionable testimony to the mere appearance, rather than actual truth, of plant immobility.[31]

Much vegetal life has the appearance of immobility due to their rootedness in the soil. This stable, fixed (but certainly *not* immobile) position on the earth's surface can be seen as doubly advantageous, particularly if there is a duality to that rootedness. It suggests a secured position from which to courageously swing to and fro in surrender, swaying flowers and rattling leaves at a tempo dictated by the breeze. All this swaying and rattling is no poetic figure of speech but actual adaptation for successful plant reproduction and defence.

The apparent immobility of plants also suggests a firm position from which to emerge, described by Elizabeth Grosz as the beginning: a chaotic and unpredictable movement of forces. Grosz develops the idea that sexual selection, the consequence of sexual difference, is the source of endless generation but is also the source of indeterminate taste, pleasure and sensation.[32] I would like to extend this idea of a beginning to the plant world. Plants' capacity to use both sexual and self-production generates multiple alternatives. Different types of sexual selection and production increase reproductive success. Plants such as the water lily adopt both types of reproduction mechanisms; moving between the two means to achieve maximum outcomes. In several species of water lilies, the female and male functions overlap, enabling self-pollination.

Gagliano explains that the vast majority of species, however, have bisexual flowers with female and male functions more or less separated in time, where

---

31   Gibson, P. and Gagliano, M. 'The Feminist Plant: Changing Relations with the Water Lily.' *Ethics and the Environment*, Vol 22, Issue 2, 2017.

32   Elizabeth Grosz, *Chaos, Territory, Art,* New York: Columbia University Press, 2008, 6.

flowers open one day in the receptive female phase to then close and reopen the following day in the male phase.[33] In some water lily species, the stigma (the female part of the flower) becomes hidden by bending itself over the center of the flower or by being covered by the stamens (the pollen-producing male part of the flower), which bend over it.[34] All this movement is agile and adaptive. To move is to think, to experience, to learn.[35]

So the act of moving and the facility of mobility stems from the 'beginning' of life. As Grosz says, sexual selection, a convention of evolution, also opens up a body of sensory activities such as 'taste and pleasure.'[36] Here, Grosz is referring to conditions associated with the sublime. However, I prefer to redirect attention to plant biology as ready-made models for better living, rather than reverting to transcendental mechanisms that remove humans (subjects) from nature (objects). Instead, I hope to suggest that water lilies and their processes of sexual or natural selection are open and emergent: they are beginnings and becomings.

And all this movement in space is also done in good time because plants know the importance of aligning their internal circadian clocks with external environmental signals. Showing a remarkable ability to adjust the speed of starch consumption in response to changes in day length, plants are able to maintain efficient photosynthesis during the day as well as optimal use of reserves during the night.[37] Clearly, time-keeping is crucial to the immediacy of plant survival, but also to its reproductive success, to ensure the continuation of the species. The opening and closing time of flowers, for instance, varies between species and has coevolved in synchrony with the presence and activities of animals. Among the water lilies, some species bloom exclusively by day and others by night. Day-blooming species are usually characterised by brightly coloured flowers that are sought out as a pollen and nectar source by insects such as bees and flies. These flowers are at most only moderately fragrant.

---

33   J.H. Wiersema, 'Reproductive biology of *Nymphaea* (Nymphaeaceae),' *Annals of the Missouri Botanical Garden* 75, 1988: 795–804.

34   G. Hirthe, S. Porembski, 'Pollination of *Nymphaea lotus* (Nymphaeaceae) by rhinoceros beetles and bees in the northeastern Ivory Coast,' *Plant Biology* 5, 2003: 670–676.

35   Monica Gagliano, co-written journal paper with Prudence Gibson, under peer review, pending.

36   Elizabeth Grosz, *Chaos, Territory, Art,* New York: Columbia University Press, 2008, 6.

37   A. Graf, A.M. Smith, 'Starch and the clock: the dark side of plant productivity,' *Trends in Plant Science* 16, 2011: 169–175.

Plant Protests and Smell

In terms of the fragrance of plant life and the sensory and phenomenological experience of plant olfaction, I must include a short discussion of Australian artist Cat Jones' work. Although the 2017 performance was not specifically on the water lily, its olfactory sensations were intense and vastly connected to the concept of a becoming-woman, a feminist and a political protest.

As part of the Sydney Festival in January 2017, Cat Jones created an experience room at Carriageworks Sydney (see figure 8). In the space were five tables with upturned cups. Each table had a theme – landscape, competition, democracy, extravagance, resistance – and there were headphones to listen to stories told by various Sydney characters, talking about how they relate to the idea of scent in Sydney. These stories related to the themes as well as olfactory memories and how the two interrelate.

So the experience for a participant, such as myself, was to wander and sit at the tables, smell the aromas Cat had made and infused in the upturned cups and listen to stories. Afterwards, I was given the chance to talk with a roaming artist-crew member who helped me formulate my own scent of Sydney and to talk about why those smells are important to me.

After being interviewed by the artist/crew member – who was also a friend, Sumugan Sivanesan – it turns out that the scents I associate with Sydney are storm-driven seaweed rotting on the beach, mould growing on garden soil, newly cut grass, the burning off from the incinerator in our back yard (where my dad burned off the garden refuse) and dead rats. I didn't intend my overall Sydney scent to be quite so fetid and manky but that's how the story of the smell of Sydney evolved for me.

The reason I liked Jones' format was because it occurred to me that the scents I had identified were all related to the artist's fifth theme of 'resistance.' Seaweed expelled from the ocean and ruining the perfect crescent of the beach; rats being killed by human-laid poisons but leaving a fetid stench far worse than their scampering could ever be; the white cobwebs of mould creeping across the soil in my back garden. These are all examples of defiance, resilience, resistance to order, resistance to human control and our desire for pristine perfection.

These smells of biotic matter that I associate with the place of my birth and my home city are plant-based. The olfactory memory of plants is strong, but their defiance of humans is stronger still. They persevere and they move, they linger and they defy.

Plant and Female Mobility

To move from fragrance, one kind of vegetal elusiveness, back to the mobility of plants, another kind of elusiveness, is to keep mobility high. Pigeon-holing plant life into non-mobility (as a representation of the supposed vegetal inability to make cognitive decisions or enact a will) is to misunderstand the concepts of nature's workings. Although much plant life is made static in their earthy position, nevertheless root systems can move for many kilometres to reach water or minerals. It is also a misreading of the ecology of vegetal matter. Bees, beetles and moths move to pollinate and then, mostly, fly away.

So the mobility of plants may appear to be limited, that is, by virtue of mostly being rooted in the ground and not being able to 'run away' from prey. However, the rapid and effective communication and movement of information within the plant itself represents an alternative form of mobility. As Aristotle says in *De Anima*, plants exhibit three of the four types of movement. They alter their state, grow and decay. The only mobility it does not have is changing its position.[38] To define mobility only in terms of a changed position in space strikes me as more limiting than being rooted in the soil. In addition, water lilies defy this fourth type of movement by being able to change position in space, via floating flowers across water. This, then, effectively places the water lily in a position of higher relevance than previously thought.

Gender-Neutrality

Plants function without organs, without a brain, yet they survive aggressive attacks, extreme weather conditions and food shortages. These, surely, are attributes of a species to pay more attention to. Plants have been associated with many gender-specific and pre-feminist narratives. Likewise, women have regularly been associated with the blooming of flowers and the blossoming of buds – the genitalia of flowers as being female with the male stamen rising up out of it.

This desire to eschew gendered preoccupations, for instance reconsidering the water lily as sexualised female deity, provokes a context within nature-culture that requires a 'becoming-other.' Can we follow the true spirit of vegetal life (its self-reproduction and interspecies reproductive assistants – beetles,

---

38   Michael Marder, *The Philosopher's Plant,* New York: Columbia 2014, 20.

moths and bees) and disrupt the notion of earth as reductively female and replace it with fluidly female? Can we reclaim nature and plant life as a queering
place, where gender is as irrelevant as the human, but difference is allowed?
Claire Colebrook engages with queer theory as it relates to logics of survival.
This survivalist theorising is important when discussing the specific qualities
of vegetal life to grow, decay, revive, and become. She writes against a generative model of life and instead discusses whether queer theory reflects on being
queer or suggests that the changing nature of being queer affects the way we
theorise.[39]

I would like to extend this point to posit that the emergent possibilities of
plant life and their hidden skillsets affect the way we theorise on nature. Colebrook's Queer Vitalism proposes that what life ought to be must emerge from
what life is. The 'self as it is formed in the social unit of the family (with the self,
taking on either male or female norms) fail to account for the emergence of
the self and the genesis of the family.'[40] In contrast, Colebrook maintains that
passive vitalism is queer in its difference and distance from constituted images
of life as fruitful and generative and humanly organized. Colebrook believes
this passive vitalism has implications for aesthetics in terms of the creation
of monuments rather than work. The body emerges and is formed through
encounters with the sensual.

Community

To subvert old habits, to show audiences the potential for change, is the key
to new philosophical thinking surrounding vegetal life. The concept of community focuses on the communication, growing patterns and rhythmic being
of the water lily as a case study for how we can learn from vegetal life. The
water lily is female but fluidly female and communally female. It is a sexuate
difference that is on the brink of change, rather than contained, discrete and
bound.

Donna Haraway quotes Don Ihde in her book *When Species Meet*. Ihde says
'In this interconnection of embodied being and environing world, what happens in the interface is what is important.'[41] Haraway proceeds to investigate
the intersections of non-human marine animals, human marine scientists,
cameras and National Geographic and television documentaries. Her point

---

39   Claire Colebrook, 'Queer Vitalism,' in *Sex After Life*, Open Humanities Press, 2014, 236.
40   Ibid, 101.
41   Donna Haraway, *When Species Meet,* Minneapolis: University of Minnesota Press,
      2008, 249.

is that these experiences and things are both inter-related and compounded (both a composite and an enclosure). They are parts and wholes that are conjoined and separate at the same time. The water lily community is similar. It can reproduce alone or together, it can move and remain stationary. The research and writing into the water lily community – the mythological significance of water lilies as feminine spirits and Luce Irigaray's writing on water lilies –are wholes and parts of wholes, communities and individuals that gather as part of a community.

As Deleuze/Guattari said, 'Follow the plants: you start by delimiting a first line consisting of circles of convergence around successive singularities; then you see whether inside that line new circles of convergence establish themselves, with new points located outside the limits and in other directions ... in nature, roots are taproots with a more multiple, lateral, and circular system of ramification, rather than a dichotomous one.'[42] Humans see the single flower, the single tree and, based on our conventional reliance on nature for our spiritual and contemplative meaning, cannot see beyond that singularity. The subterranean language of plants is a series of chemical emissions and receptions. Deleuze and Guattari's concept that language can be broken down into internal structural elements and is not fundamentally different from a search for roots strikes an accord with what new science now shows – that there is rhizomic assemblage multiplicity rather than unity or units of measure.[43]

The multiple, then, is not a single rose growing on its own, nor is it a single woman gazing upon that one rose. The multiple is the community of plants, its ecosystem of like species among other species. Fungi help to send messages from root to root. Beetles and moths help pollinate at night. Various roots from various plants search for nutrients and water and communicate their findings. This, then, is not a male patriarchal or hierarchical system of politics. It is an aggregated and equalized community.

There is a desire among plants to survive, to flourish. Like the concept of becoming woman, this could be considered a desire to communicate and contribute to the lives of our neighbours as much as ourselves. Could this model of vegetal behaviour be worth more than we previously thought? Perhaps more urgently, does this desire to flourish change our preconceptions of plants as inert, immobile and incapable of thought processes or decisions-making? If so, what effect does this epiphany have for culture, for communities, for women?

---

42   Felix Guattari and Gilles Deleuze, A Thousand Plateaus, Minneapolis: Minnesota Press, 1987, 26, 34.

43   Ibid, 29.

For the feminist woman, a plant that self-reproduces, works as a team member and is mobile beyond human definitions could be a revelation of political force. Plants are the becoming of the becoming of each other.

A quality of being among a community is the ability to care. From insemination to dissemination, animal to plant, the reproduction of plants relies on winds and inter-species.[44] From the structure of an animal co-species, from the words of Donna Haraway, comes a plant co-species test. Where Haraway speaks of seeing again, holding in regard, to esteem and pay attention, she is speaking of humanity's relationship with dogs. Here I apply that respect to plant species, particularly water lily.[45]

Her concept of becoming woman is that every woman is an 'actualisation of the potentiality to be female.' For a group of humans, this means the energy lies in the force of that potentiality. Nothing has arrived, it is in flux. *The plant contract* is a radical re-thinking and a celebration of what differences there might be. Beneath these stories of artworks and of women is a tacit agreement, a contract, to pay more attention to the plant stories within each moment.

---

44   Elaine Miller, *The Vegetative Soul: From Philosophy of Nature to Subjectivity in the Feminine*, Albany: SUNY Press, 2002, 184.

45   Donna Haraway, *When Species Meet*, Minneapolis: University of Minnesota Press, 2008.

Ungrounding Plant Life: The After-Effects

By growing beyond Leviathan, past a critical mass, the collective moves up from monster to sea, while falling from the living to the inanimate, whether natural or constructed. Yes, the megalopolises are becoming physical variables: they neither think nor graze, they weigh. Thus the prince, formerly a shepherd of beasts, will have to turn to the physical sciences and become a helmsman or cybernetician.[1]

This *plant contract* leans heavily on the work of Michel Serres and his natural contract. His emphasis on lost contact with nature and lost contact with one another strikes at my heart. Serres' herd of Leviathans is lost. There is no new prince, no leader to direct us through the quagmire of climate change and lost connections with nature. All we have left are the plants, and story and art. The plant contract is more than an urging to take better care of how we see the vegetal world. It is also a tale, a story, a parable, a fable, a yarn. There is a secret potion, a magical kernel at the heart of the myth. And that is the plant. The plant is the only prince, the plant and all its incarnations.

Deterring Atropy

It is hard not to be in a state of exultant praise for the plant. Vegetal science increasingly illuminates the value of plant life. There are qualities to plant life, however, that are atrocious, dangerous, parasitic, painful, competitive, deadly and ... illegal. They are magic. Think of the *Atropa belladonna*, deadly nightshade which was often included in a brew of mandrake, hemlock, henbane and opium, as an early anaesthetic. In 1915 Professor Henry Walters declared that plants were capable of love and memory and that deadly nightshade was full of hatred, hence its purposeful, poisonous ways.[2]

---

1   Michel Serres, *The Natural Contract*, Ann Arbor: University of Michigan Press, 1995, 20.
2   Stewart, Amy. *Wicked Plants: The Weed that Killed Lincoln's mother and Other Botanical Atroci-ties*, Algonquin Books, Chapel Hill, 2009, 31.

It is important, here, to acknowledge the more malevolent side of plant life as well as its positive attributes. Plants crowd each other out, they steal nutrients. Yet there is much language in the public domain that perceives these actions as 'sharing' and 'teamwork.' For instance, Peter Wohlleben wrote a forest book about the benevolence of plants to one another.[3] He starts his story via an anecdote about finding some wood just beneath the surface of the forest floor – which was the last of an ancient tree stump, still alive because its colour was chlorophyll green. He concluded, without substantive data or proof, that the surrounding beeches were sharing their nutrients (sugar), processed through the photosynthesis of their leaves, with this tree stump. Was the stump stealing or were the plants giving?

This book has focused on the difficulties of these non-human interrelations. Humans have no access to knowing which plant element was giving and which was being stolen from – all we know is the movement of the nutrients. The disparity between science and the humanities is very clear in this instance. These distinct disciplines, however, intercept. And this chapter focuses on some of those interception points. It focuses on how plant matter has been used for magic and medicine. And it focuses on how art aesthetics mediates new human relations with nature (plants), in particular in the medicinal artwork of Janet Laurence. These are both situated in the aesthetics of care or cure. In this instance, I will present both the kindly and the wicked side of plants and our human mirroring of those two affective elements in art.

The Magic of Secret Herbs

A plant contract is created from the magic of plants. To think we can understand plants is only to understand our human relations with them. Outside that inter-relation, we have no further access. We can grow plants and harvest them, mix them and cook them. This is an interference that creates a change in those plant species. It is a human manipulation for human purposes. Witchcraft and ancient medicine is also a process of using plants and changing them, for the purpose of changing the physical state of a patient. Yet again, like agriculture and farming and cooking, this is a relation with plants that is utilitarian. However, there has always been greater respect for plants, by witches and shamans, than surrounding other forms of plant utility. It is that respect for plants in witchcraft and other sorcery that is celebrated here. This celebratory story of plant-change is a clause of the *plant contract*. This includes a respect

---

3   Wohlleben, Peter. *The Hidden Life of Trees*, Back Inc, Sydney, 2015, 2.

for the independent agency of plants and their capacity to create change in other species.

The history of independent malevolence/benevolence of plants is well documented in medicine and alternative healing. From shamanism to Western magic, plants have a lineage of altering human states. Isaac Newton was intrigued by the idea of discovering the Elixir of Life, the secret to immortality. His working papers were splashed with alchemical mixtures that suggested he was experimenting to find the Elixir of Life, the magical potion of infinity.[4] According to Jamie James, Newton was an occultist at the turn of the century. James suggests he studied 'the fragments of magical books attributed to Hermes Trismegistus, and translated one of them, The Emerald Tablet. A basic text of alchemy ...'[5] Newton worked on two treatises on prophecy,[6] in which he interpreted parts of the Bible and the hieroglyphic language of the Egyptians as prophecies to lead him forwards. In an attempt to make sense of heaven and earth, sun and moon, fire and meteors, animals and minerals, man and beast; he came to mystical qualities, prophets and what must have been a bewildering concern that chasing the philosopher's stone might be seen as a blasphemous act.

The plant and mineral extracts that Newton was experimenting with were an extension of widespread medical use of plants for anaesthetic or treating fevers. In the collection of the National Museum of Australia, for instance, is an 18th century medicine chest that belonged to an Australian family, aptly named the Faithfulls. The concept of an 18th century medicine chest endures today. Many of our tinctures and vitamins are plant-based or natural mineral-based. Contemporary society's obsession with taking vitamins and other tinctures has been described by doctors as doing no more than creating expensive urine. In other words, the efficacy of much herbal medicine is dubious.

The Faithfull Family Chest comprised Dover's Powder – an ipecac rhizome and root syrup; a vial of James' powder – an oxide of antimony (natural mineral) which cured gout and scurvy fevers; laudanum – a tincture of opium poppy which worked as a narcotic and was an addictive analgesic; and a vial of paragoric elixir – a camphorated opium that relieved diarrhoea.

The relevance of the medicine chest is how plant and mineral biotic matter was used to treat and cure patients, and how cures and elixirs are made today – controlled, distributed and marketed. It is also interesting to consider

4   Jamie James, "W.B. Yeats: Magus," *Lapham's Quarterly*, Summer (2012), accessed August 12, 2013, http://www.laphamsquarterly.org/essays/wb-yeats-magus.

5   Ibid.

6   Newton, Isaac. *Two Incomplete Treatises*. Bettenham, London, 1745.

the aesthetics of the medicine chest. These bottles were meant for keeping and re-using, whereas all our contemporary medicines are bottled in plastic and unsustainable packaging.

The efficacy of these plant-based drugs has much to do with faith in their curative possibilities – the placebo effect. How medicine and cures are given is as important as the components of the elixir itself. In his paper 'On Magical Language,' Stephen Muecke outlines four utterances of magical speech. Firstly, there is the exorcism, where bad objects disappear. Secondly, there is the imprecation where a bad object appears. Thirdly, there is the curse of commination (the threat of divine vengeance), where a good object disappears. Last is the blessing or conjuration, where a good object appears. As he says of his adaptation of Todorov's 1978 *Le Discours de la Magie*, 'this classification helps us distinguish between two opposed forms of magical language: the spell and the prayer.'[7]

The spell, like the prayer, is a powerful force. It is a device of suggestive language with an intention to change a relationship through the vitality of an object. It is a summoning of force. The vital object might be a good luck charm. Alternatively, there might be a turn away from goodness by using voodoo or a poisonous potion. The spell can invoke a positive or negative incantation. Anthropologist Michael Taussig refers to the impure sacred, the idea of primitive religion working against the goodness of the church. He tells of how the 'forces conjured by the sorcerer, and the blood issuing from the genital organs of women ... inspired men with fear.'[8] The use of vegetal-related objects to create change is necessary. This might also be parchment with the language of the spell written upon it. Or it might be the herbs used for potions. Where is the force held? With the humans who use the plants or with the plants themselves?

Ecuador is famous for its traditional medical practitioners – curanderos – particularly in the highlands. Normal plants are used as elements in traditional medicines but there are also magical plants which have special properties.[9] The biodiversity of the plants in the upper and lower rainforest aids the exoticism that has subjects paying for *limpiezas* or cleansing rituals. Plants are used for casting spells for social problems, but these curanderos differ from the shamans who participate in the use of plants for altered consciousness and

7   Stephen Muecke, 'On Magical Language; Multimodality and the Power to Change Things,' in Margit Bock and Norbert Pachler, eds., *Multimodlity and Social Semiosis: Communication, Meaning-Making, and Learning in the Work of Gunther Kress* (New York: Routledge, 2013), 91.

8   Michael Taussig, *The Nervous System* (New York: Routledge, 1992), 114.

9   Anthony Cavender, Manuel Albin. "The use of magical plants by curanderos in the Ecuador highlands." *Journal of Ethnobiology and* Ethnomedicine 5, 3, 2009.

spirit awakenings. *Mal air* or bad wind is reported as the most difficult illness to cure in Ecuador and it is interesting to note that the most common shared characteristic of plants chosen as cures is their odoriferousness ... their scent.[10] The odour draws the toxins or bad air from the body.

Many Westerners are wary and sceptical of indigenous or eastern magic. It is perceived as non-normative, sometimes associated with quasi-religious fervour and unsubstantiated superstition. In her essay *Queer Performativity*,[11] Karen Barad explores acts of nature as acts against nature and seeing ourselves as always part of nature. If I read 'nature' as the world, and the actor as the specifically human actor, then magic might be seen as a 'queer act,' an act against inequality. Barad writes of Nazi ss elite henchman Heinrich Himmler's delousing campaign – Jews were deemed unclean and the genocide was human against non-human.[12] The early witch-hunts operated in a similar way.[13] Witches were non-human and therefore to be exterminated. Barad writes, 'the "posthumanist" point is not to blur the boundaries between human and non-human ... but rather to understand the materialising effects of particular ways of drawing boundaries between human and nonhumans.' Her point is that agency and abjection come into play when performative accounts are made of the non-human.[14]

The political and vegetal act of dispensing magic spells, charms and potions is a weighty responsibility. For instance, by giving Shakespeare's Juliet the sleeping potion to feign death, Friar Lawrence causes a series of unfortunate events, not to mention his mobilising of political and personal tragedy. Relying so heavily upon faith and mutual respect, the causal act of plant-based potion and charm-giving has a multitude of dangers and potentially negative reactions: misinterpretation, disappointment, anger. The worst must be the burden of accountability: what happens if an audience member or receiver takes action as a result of the stimulus of the potion or charm, which was not expected or is harmful to the receiver or others?

In his book *The Magic of the State*, Michael Taussig created a ficto-critical tale of fieldwork in the mountains of Venezuela, where pilgrims came and were possessed by spirits.[15] The rituals of magic and spirit possession were meant as an elaborate metaphor for the authority of the state, its hierarchy and

---

10   Ibid 7.
11   Barad, Karen. "Nature's Queer Performativity." *Qui Parle* 19, 2 (2011): 121–158.
12   Ibid, 122.
13   Kramer, *Malleus Maleficarum*. Speyer, Germany 1487.
14   Barad, Karen. "Queer Performativity," 123–124.
15   Taussig, Michael. *The Magic of the State*, Routledge, New York, 1997.

stratification. 'This is an anthropology not of the poor and powerless, but of the state as a reified entity, lusting in its spirited magnificence, hungry for soul stuff,' he writes.[16] The dramatisation of the spirit possession act and other sorcery was as constructed as the pomp and ceremony of public political events and the official authorial voice.

In this way, Taussig is interested in the difference between the reality of the state and its impure and untruthful mask. In other words, the state is a false idol and not the sum of its parts. Instead, it is a conglomeration of parts which give the appearance of a united authority. He referred to Philip Abrams' concept of the State as being not the reality behind the mask of politics, but the mask that prevents us from seeing political reality.[17] It is phenomenology that seeks to remove the mask, pull away the veil and uncover the reality behind the surface or the face. The partaking of vegetal brews is a similar phenomenological act, a sensory and experiential action.

Desire for a Sacred Plant

Plants have a sensory and experiential effect on humans. We imbibe plants and touch plants, use them to incite vomiting and pick their flowers for aesthetic pleasure. This is all human action with plants, rather than plant action with the human. If nature has conventionally been a background, a grounded base against which all action occurs, then this concept leaves nature without agency and without an independence from the human. But what if this naturalistic view was erroneous? What if the natural world – the forest, the bush, the rainforest, the jungle – is the action and humans are the mere witnesses, more inert and incapable than we have previously thought?

In the 2016 black and white film *Embrace of the Serpent*, Colombian director Ciro Guerra follows the travels of two real-life European explorers into the Amazon (based on expedition journals in 1909 and the 1940s). The white men are guided by shaman Karamakate (the last of his tribe, and against his better judgement) to find an elusive healing plant called the yakruna. This sacred and secret plant has curing properties. Meanwhile, the first European explorer, Theodore Koch-Grunberg is sick with fever and treated with 'the sun's semen' by Karamakate. This white powder, applied in blown doses up the nostrils, may be a cocaine-like drug substance. The yakruna, eventually sampled by the

---

16   Levi Strauss, "Interview with Michael Taussig; The Magic of the State," *Cabinet magazine*, 18, Summer (2005), 3.

17   Taussig, Michael. *The Nervous System*, Routledge, New York, 1992, 113.

second explorer, Richard Evans Schultes, seems to act on him like a hallucino-
genic. In fact, there are references to Schultes taking caapi in the film, a privilege that has to be earned. Caapi is one of the plant extracts that makes up the
brew ayahuasca, a popular mind-altering drug taken by tourists in Peru.

Caapi and ayahuasca (combining vine and leaf extracts) were originally
used as spiritual, shamanistic sacraments – a means of curing and opening up
paths of knowledge. The psychedelic visual and auditory stimulations of the
drug have attracted Westerners and this has created a tourism market for those
craving 'experience.' The protection of the plant as traditional knowledge has
more recently stirred up questions of intellectual property.[18]

Back to the search for the yakruna plant. The *Embrace of the Serpent* film
accentuates the disjuncture between plant-drug as cure and as cultural object.
More important is the notion that the jungle protects and conceals its sacred
plants. Why sacred? What makes caapi non-spiritual and yet the yakruna or
ayahuasca sacred? Perhaps the sacredness was originally associated with the
difficulty of its procurement. In the film, the plant is a tree that blooms with
a magnolia or orchid-like white flower. There are few sources left (in this tale)
and the shaman Karamakate has burned one cluster of the plant when he sees
a community abusing its uses in a non-scared context, that is, to get off their
heads with delirium. The ayahuasca brew has anecdotal associated stories too,
of the complexity of procurement. The root of the vine is mixed with the leaf
of a same-species bush. Indigenous Peruvians are said to be able to hear the
plants communicating, this being the means of finding the active ingredients
for the brew.

Sacredness assumes spiritual qualities (above the average), and an ordained
property (chosen by a community as having protected status) – and it also assumes that only the few, the sovereign, have the right to reveal it, to impart
the knowledge, to prescribe it. In the film, Schultes begs permission to try the
yakruna and, failing that, the caapi. The shaman denies him access due to his
inability to believe, to have faith. So those things that are made sacred become
the most desired because they are denied or made difficult to achieve. What
does this say about the forces of desire amongst all communities, amongst all
people?

---

18   Tupper, Kenneth (January 2009). "Ayahuasca Healing Beyond the Amazon: The Global-
      ization of a Traditional Indigenous Entheogenic Practice." *Global Networks: A Journal of
      Transnational Affairs.* 9 (1): 117–136; McKenna DJ, Callaway JC, Grob CS (1998). "The scien-
      tific investigation of ayahuasca: A review of past and current research." *The Heffter Review
      of Psychedelic Research.* 1: 65–77.

Richard Doyle focuses on plant-based drugs, such as ayahuasca, as living organisms, and as technologies that might explore what it is that lies beyond thought.[19] His long-term asthma was cured as the result of some shamanistic care and the use of ayahuasca. Dale Pendell also follows the drug amphetamine and its 'tendency to monomania.'[20] A stimulant, it was a drug first produced by German chemist Edelsno in 1887 and was included in emergency kits for soldiers' survival. Adolph Hitler is reported to have been injected with meth-amphetamines daily.[21]

Amy Stewart writes that the sci-fi writer William Burroughs drank aya-huasca tea. Paul Simon, Sting and American nature writer Theroux did too. So much creativity, so much plant-induced intoxication! As Stewart reminds us, chacruna contains the psychoactive drug dimethyltryptamine that has to be activated by another drug (in this case the bark of the woody ayahuasca vine). People have reported horrendous hallucinations and vomiting but that it also can cure addiction and depression.[22]

In 1895 Sigmund Freud wrote that 'a cocainization of the left nostril has helped me to an amazing extend.'[23] Coca leaves as a stimulant have been used since 3000 BC, placed between cheek and gum. Italian doctor Paolo Man-tegazza, who prescribed coca leaves, wrote: 'I sneered at the poor mortals con-demned to live in this valley of tears while I, carried on the wings of two leaves of coca, went flying through the spaces of 77,438 words, each more splendid than the one before ...'[24] All writers have experienced that sense of writing well, only to wake up the next day, approach the text, and see that those words were not quite so great as originally thought. But perhaps the alkaloid cocaine helps with that disappointing phenomenon.

The Great Ungrounding

The exaltation of plants and their capacity to alter experience via elixirs and cures ties into the second idea of this chapter – ungrounding and the

---

19   Doyle, Richard. *Darwin's Pharmacy: Sex, Plants and the Evolution of the Noosphere,* Wash-ington: University of Washington Press, 2011.

20   Pendell, Dale. *Pharmako Dynamis: Stimulating Plants, Potions and Herbcraft,* North Atlan-tic Books, Berkeley 2010.

21   Ibid, 150.

22   Stewart, Amy. *Wicked Plants: The Weed that Killed Lincoln's mother and Other Botanical Atrocities,* Algonquin Books, Chapel Hill, 2009, 8.

23   Ibid, 21.

24   Ibid 22.

Anthrobscene (a term coined by Jussi Parikka in his paper of the same title, which refers to the obscene layer of the earth's crust which is now full of disused metals and computer and phone refuse) as it relates to artist Janet Laurence's medicinal works. But there are problems to deal with first. By ordaining and sanctifying plants and trees, such as the fictional yakruna, there is the danger that our theorising has slipped back into romantic idealism. By setting plants apart as scared and ordained, the worship of nature is still steeped in a purpose that serves the human. The stories of the shaman's yakruna, however, tells us other stories as well. It reveals human reliance on plants and the unwitting ignorance of white people who are unable to believe in the power of the vegetal. It also reveals how quickly even indigenous people can lose the knowledge due to the Anthrobscene, that vile last epoch of human geological time, when the surface of the earth has become foul and poisonous due to human use of plastics, metals and disused technology (discussed later in this chapter). The Anthrobscene theory helps us to unground nature and see what 'we' have done. But the Anthrobscene is not just a geological stratum, it also infiltrates the atmosphere above the earth too.

The process of ungrounding is a mighty task: to alter our concept of the earth as a place 'for humans alone.' It requires a shift in our perception of nature. It also requires a warning against falling back on indigenous humans to answer all our prayers, at a time when so-called civilised cultures have defiled the very culture they now seek to return to. We also need to beware of combatting the grounding of nature as mere backdrop to human action by elevating the status of plants as icons or deities. This does not solve the problem. The only way to redress the balances of humans within the natural world, and the natural world within the human, is to create a structure of thinking that is aggregated and circular, rather than hierarchical and chronological.

Magical Plants Ungrounded for Art

Australian artist Janet Laurence is an important figure to introduce within this ungrounding and plant drug discussion. She toyed with the hallucinatory, stimulant and intoxicating effects of various natural plants in her artwork *The Elixir Bar* 2005. It was a concept of plant specimens and potions as cures, where she collaborated with local Japanese villagers and botanists to grow herbs and harvest them for medicinal tinctures. So, for the Echigo-Tsumari Art Triennial, for which *The Elixir Bar* was developed, Laurence was invited to Japan to choose a site for her work. During her time in Japan, in the lead-up to the work, *The Elixir Bar*, Laurence became concerned with the question of how to house nature. It was an intense collaboration with local Japanese farmers to create a permanent installation. Laurence was given the use of a traditional wooden house. There was a connection to the immediate environment – along with the wonderful wooden structure perched on a mounded hill, nestled near tall evergreen trees.

As part of her research, Laurence was able to accompany local plant experts on expeditions to the mountains to collect herbs and plants. Her personal association to the countryside and people became very strong over that time. She planted a specimen garden outside the traditional house that was used for the installation. Inside the *Elixir Bar*, enormous glass jars of plant extracts – nuts, herbs and seeds steeped in *shochu* (distilled from barley or rice) were presented and labelled as part of a laboratory or apothecary. This work ties into the history of medicinal plant activity, including the cultivation and harvesting of plant matter. The direct contact with the plant source is typical of Janet Laurence's process. She does not just create immersive artworks, her artistic process and praxis is immersive. Products and usage derived from plants are surely a process of hybridity between humans and plants. We ingest these plant extracts as we sip at herbal tinctures. We swallow plant after plant after plant. How can we not pay attention to this matter that forms a massive part of our dietary requirements? Janet Laurence makes aesthetic this basic and fundamental act of humans: the imbibing of plants for medicinal purposes and to explore the tastes of different plants.

The *Elixir Bar* is a permanent work and is now cared for by the custodians of the village. *The Elixir Bar* represents wonder, exchange, collaboration and shared knowledges of traditional medicine and nature. But it also sounds a strong message of alteration, transmutation and biological change due to plant extracts. This work was a major collaboration, an extended project, and it also was an alchemical act. The procurement of plants and the use of them as cures and medicinal potentials were at its core. In this way, it continued Laurence's long-held fascination with change and transformation.[25]

Laurence uproots or ungrounds vegetal matter for her exhibition installations. She also conjures new experiences. She could be described as a contemporary witch of the art world, a woman who transforms natural specimens to create a change. Irigaray says, 'For a long time, women remained closer to a world of roots. And, especially, those who have been called "the witches," who remain faithful to the natural world, the living and curative properties of which they know.'[26] My hope is to join Janet as a fellow plant witch.

25   Gibson, Prudence. *Janet Laurence: The Pharmacy of Plants*, Sydney: New South Books, 2015, 111–112.

26   Luce Irigary and Michael Marder, *Through Vegetal Being*, Columbia University Press, New York, 2016, 39.

Woodard's Ungrounding

Many of the discussed changes in perceptions of nature connect with Ben Woodard's interpretation of an "ungrounding" of the earth. Woodard analyses speculative realist philosopher Iain Hamilton Grant's writings on nature-philosopher Friedrich Schelling (1775–1854) and his concepts of the ground. Grant refers to the problem of sufficient reason in concepts of 'naturalism' as never exhausting its ground. The ground cannot be reduced. The ground does not depend on a human witness. If the earth is ungrounded then it is no longer a backdrop, it is no longer inert. According to Woodard, ungrounding is a geological digging in the soil, as a means of disrupting our perception of the planet's surface.[27] (He is referring to mining machinery and agricultural infrastructures.) Woodard draws upon Schelling, who urges for a move away from privileging the world over the earth, or the concept over physicality. In other words, the human habit of thinking about nature as an external phenomenon needs to be overturned.

Janet Laurence's BioArt deploys plants as a conceptual apparatus and her work is an example of the growing scientific and artistic interest in the neurobiology and communication capacities of plants. Her work engages with the relevance of plant life as a means of redressing the place of humanity in the world. It helps to shift the ontological significance of plants as independent and agented beings, which subsequently changes their aesthetic value. This shift towards a more equalised register of being affects our cognition and perception of plants and plant-thinking. It also affects how the functions and adaptability of plant life can inform the way we live in, and consciously perceive, the world around us.

Janet Laurence has a biocentric view of the world. Much of her work incorporates live biotic matter such as plant seedlings, tree branches, root systems and plant-related material such as seeds and soil. Her use (or re-use) of plant life in an art setting, often conducted as a 'concept of care,' suggests the lively elements of nature. Her aesthetic interpretation of the physical benevolence of the earth is important for a revised understanding of the environment as an energetic operative and as a system of information, rather than as an inert backdrop to human action. Laurence repurposes plants in her art. Her plant artworks are mobilised out of the fecund ground and into the gallery space and, as such, they address plants' ontological status and engage with theories of the grounded geological earth.

---

27   Woodard, Ben. *On an Ungrounded Earth: Towards a New Geo-Philosophy.* New York: Punctum Books, 2013, 3.

The eco-psychology of our global society (exhibiting symptoms of extinction fear and critical climate change concerns) and new discoveries in how plants operate, adapt, and sense all things inform this analysis of Janet Laurence's important plant-specific art work. Her art practice fits within the theoretical framework of the geo-philosophy of Jussi Parikka and his concept of the 'Anthrobscene,' which marks the abject results of mankind's impact on the earth's geological surface.[28] It also links with the 'plant-thinking' concepts of philosopher Michael Marder.[29] It does this in part by acknowledging the problematic anthropocentric divide between human subject and natural object – through her creation of immersive, interactive and experiential installation encounters – in order to alleviate and diffuse a conventional and divisional dyad relationship.

Laurence's interactive artworks draw attention to the marks humans have left on the earth's surface through agriculture, mining, land-clearing and urban development. The artworks highlight the issue of how new discoveries in plant science show us that plants can smell, hear, think, learn, remember and communicate.[30] This information changes our perceptions of, and relationships with, plant life and with the surface of the earth and nature more broadly. The impact of these changes is not yet fully known, but its evidence creates a more agile passage between the disciplines of art, philosophy and science because it redresses any hierarchical models of being. Artist, plant, gallery floor, writer, soil, viewer, branch and thought: these are all of equal register.

Janet Laurence has been creating and exhibiting art for over twenty years in Berlin, Paris, Asia, the US and Australia.[31] Her work reflects her perturbation over critical climate change issues and it investigates the accelerating crisis she sees in the natural environments she explores and repurposes. She has collaborated with scientists, pharmacists, herbarium scholars, landscape gardeners and botanists over the last twenty years and she has worked with the IGA exhibition organized by the Berlin Senate at the site of the former Tempelhof airport and Blumberger Dam for 2017. *Gardens of the World* was a 25 hectare section of the overall 66 hectares dedicated to artists working with garden elements, and Laurence's work was installed in this section.[32]

Laurence was invited by IGA to visit the Botanical Museum in Berlin-Dahlem, located at the Free University, to meet various academic plant experts.

---

28   Parikka, Jussi. *The Anthrobscene*. Minneapolis: University of Minnesota Press, 2014.

29   Marder, Michael. "What is Plant-thinking?" *Klesis: Revue Philosophique*, 25 (2013): 124–143.

30   Chamovitz, Daniel. *What A Plant Wants to Know*. New York: Scientific American, 2012.

31   http://www.janetlaurence.com/.

32   https://iga-berlin-2017.de/.

She made contact with Professor Matthias Melzig at the Institute of Pharmacy at the Free University Botanical Gardens, an expert in psychotropic plants, to access specialist plant knowledge. Laurence's aim for this project was 'to connect myth and magic of plants with medicine, through art.' Thus began her research into the properties of plants, their toxic qualities and transformative effects. She investigated the history of the herb garden at Dahlem, the planting of which is laid out and cultivated in the shape of a human body. She also researched the 11th century mystic, Abbess Hildegaard Bingham, reflecting a long history of research and usage of plants for medicinal purposes in Europe.

For her 2017 IGA installation, Laurence began research into psychotropic bio-science, and then created a preparatory design for a medicinal garden for her work, *Inside a Flower* (See figure 9). The dome was cool and washed with white light. Entering the pod, the visitors wandered around, having arrived at her geodesic flower structure. This geodesic (strutted dome) form was

FIGURE 9
Junet Laurence. Inside the Flower –
for IGA Kunst Berlin 2017. in
association with LAVA Berlin and
CityPlot Amsterdam. transparent
mesh, mirror, water lens, tubes,
specimens and medicinal plants.
PHOTOS BY: LESLIE RANZONI.

designed like the interior of a flower. The units of the exhibition were elaborated as cellular elements of nature made large, ungrounded from their original purpose. The audience could taste the various plants, via hanging vials and
curling medical tubing, on the way to the flower. The hallucinatory effects of
the plants might alter the visitors' experience of the flower. Viewers partici-
pated fully with the plant world as equals. This is not plants 'for us' but plants
'with us.'

The access Laurence was granted to the Botanical Garden and Botanical
Museum of Berlin's Plant Science Library during her research trip to Berlin
in 2014 provided resources and research material for the exhibition. She was
granted entry to the herbarium and particularly the Dahlem Centre of Plant
Sciences, also at the Free University, which runs across 43 ha and holds 22,000
plant species. In 1879 the herbarium was established and now has a Japanese
garden, several glasshouses, and orchid and carnivorous plant pavilions. The
artist's curiosity has long been one of empathising with plant life, creating clin-
ics for sick plants (such as her Biennale of Sydney installation *Hospital for Sick
Plants 2010*) and using the spaces of galleries and museums to form museologi-
cal associations with specimens, to create awareness of their extinction threat.

Turning towards Plants

Laurence's work, then, sits within a framework of changing perceptions of nature, particularly the history of human impact on the environment. Geophi-
losopher Jussi Parikka weaves connections between culture, morals and the
geology of the planet in his book, *A Geology of Media*.[33] He expands debates
regarding the Anthropocene, scientific cultures and technological realities
through a discussion of artists' perspectives. His method of establishing discourse surrounding pollution and mass extinction, and the ways in which artists mediate these eco-issues, has had an influential impact on this chapter,
which analyses the art work of Janet Laurence in the context of the Anthrobscene, first, followed by a discussion of her work in the context of Schelling's
'ungrounding' and then Marder's 'plant-thinking.' If we understand the Anthropocene as the collision of nature, culture and industry during the late
nineteenth century, then Parikka's incorporation of technology and geology
into the analysis of the layers of the earth, as a record of human impact, results
in the term he has coined.[34] Parikka's provocative Anthrobscene is the obscene
impact of mining the earth for resources, without due care. Parikka describes

33   Parikka, Jussi. *A Geology of Media*. Minneapolis: University of Minnesota Press, 2015.

34   Parikka, Jussi. *The Anthrobscene*. Minneapolis: University of Minnesota Press, 2014, 1–8.

the carboniferous Anthrobscene as 'the layers of photosynthesis that gradually were being used for heating and then as energy sources for manufacture in the form of fossil fuels.'[35]

This 'obscene' new stratum or top layer of geology (the result of electronic waste, climate change and energy depletion) is important to any theory regarding plant life and aesthetics. We can think of plants as philosophical enactments of nature and beauty, but also as survivors of human over-production. Their roots extend down amongst the debris of mobile phones and redundant tablets, where the digging harvester has scraped and the miners have dug their shafts. Plants' branches reach out towards the distant stars of Virgil's poem too.

Could plant art be the mediation between the historical damage wrought upon the environment and the possibilities for change in our attitude to resource usage? This 'turn towards plants' would require a move away from the historical categorisation of plants, the systemic collection, classification and taxonomy of plants, and the curating of plant specimens in storage facilities. It would, in fact, require a complete reordering of the ontological status of plants, with a concomitant receptivity to the concepts of plant learning and memory.

Jussi Parikka is conscious of the divisional collapse between subject/object, and between culture/nature, however his vision is one of the earth as a geology of media, as deep-time resources that might intermingle with the roots of trees.[36] Parikka brings attention to the underground as tangential to the ungrounded (digging into the earth, and digging as deep thought) and to the invisibility of sub-surface activity.[37] How can we consider a turn towards Laurence's relevance of plants without wondering about their sub-strata activities? Laurence's work functions as a defense of Parikka's ideas, due to her consistent placement of organic and inorganic elements next to each other in her installations and her ability to bring the memory of the deep earth inside the gallery space. *Inside a Flower* (2017) exhibited real biotic plant matter and constructed versions of a flower alongside each other. In *Hospital for Sick Plants* (2010), Laurence placed plants and plant matter in a white medical tent (like a glasshouse) in the Sydney Botanical Gardens and connected them to medical tubing or wrapped them in gauze, placing them alongside beakers and petri dishes. Science + Nature = Art. Aside from the obvious interpretation of nature requiring resuscitation, the work also connects with Parikka's issues of nature and culture being inter-dependent. This work might be considered as a re-privileging of the human over the non-human, as it refers to what plants

---

35   Ibid 17.

36   Ibid.

37   Parikka, Jussi. *The Anthrobscene*. Minneapolis: University of Minnesota Press, 2014, 8.

can do for us. However, it is also a plaintive call regarding what we can do for plants. This model is an aggregated and inter-species network of activity.

Attempts at a less human-focused view and a tolerance for the likelihood of expanded plant properties are evident in Laurence's IGA Berlin work. *Inside a Flower* was a hedonistic and alchemical tour of the workings of a plant. Magic and medicine meet where the human participant seeks to be transported by the agented power of the plant: to become one with the flower. This is plants as they really are: a source of information, a system of complex networked and aggregated action. Photosynthesis, rhizomic activity, transfiguring neural properties and the lengthy process of growth extending beyond the lifespan of the human (especially in the case of trees) are elements or conditions from which we may learn. Parikka's media materialism is enacted by Laurence, in terms of re-purposing medical equipment and enacting a system of care, and refers to the past but also addresses the way we engage with plant life and the soil beneath our feet.

Ben Woodard's Ungrounding

Philosopher Ben Woodard writes, 'The digging or ungrounding of the earth is often tied to thought.'[38] The concept of the ground refers to a transcendental geology, a production of matter relating to deep time. However, in an artistic context, the 'ground' refers to the base of the artwork. The painting canvas, the piece of paper and the sculptor's stone all constitute types of ground. The ground in art has suffered a similar conceit as the ground in nature, in that they are considered no more than a background. Ungrounding or regrounding a concept of geological time has implications for how we understand the relationship between human and nature and art. The world need not be privileged over the earth. Mankind need not be privileged over plants. This, then, refers to Janet Laurence's movement of plant and earthy matter out of the ground, literally and conceptually. By doing so, she contributes to the possibility of reconsidering our human place in the world.

In fact, we exist within a grounded nature and we are also gravitationally grounded to the earth. Laurence's artwork resonates with these ideas as she, conversely, hauls plants out of the earth and presents them in the gallery spaces. She also resuscitates them – keeps them growing in artificial spaces,

---

38   Woodard, Ben. *On an Ungrounded Earth: Towards a New Geo-Philosophy.* New York: Punctum Books, 2013, 8.

re-performing their functions and caring for them over suspended periods of time in an aesthetic context.

Laurence's work ties into an ungrounding in terms of redressing our perception of the earth's surface (and the way plants puncture both below and above) and all the objects and detritus and mismanagement that contribute to the earth's temporal geological strata. When her plants are removed from their natural environment and regenerated in artificial spaces, this distinctly draws attention to how and why we have seen plants as immobile. Instead of that paralysis, Laurence mobilises plants by moving them to the gallery space, to new outdoor sites, to gallery storage/collection places. The accepted axiom that plants are incapable of sentience or intelligence or mobility can change, in the context of geo-philosophy, through Laurence's encouragement of audiences to interact with her plants directly via tinctures and unorthodox museum displays and installations.[39]

To relate Laurence's *Inside a Flower* to an ungrounding, I use Woodard's discussion of an external and putrefied ungrounding. In his book *On an Ungrounded Earth*, this concept of shifting earth perceptions eventually leads him into the nebulous world of science fiction, a common habit among theorists who shave close to the principles of Speculative Realism theories. Woodard acknowledges that Iain Hamilton Grant has brought renewed attention to the ungrounding concepts of Schelling. In Woodard's discussion, the earth is not just the world – the earth thinks through its inhabitants, it does not exist for our delectation: it is the "unthinged."[40]

Geophilosophy extends to aesthetics. In a complementary fashion to these critical philosophical concepts, Laurence's artwork, in an illustrative capacity, creates a more concrete and material version of an ungrounding. It makes reference to an archaeology of things as fetishized objects (such as plant specimens and laboratory apparatus as artwork components) and it refers to human interaction with plant specimens. Her art mediates or punctures the problems of human privileging by serving as an inter-connecting relation. Plants connect geology with atmosphere and Laurence connects art with science.

Growth is an important characteristic of plant life. In an unpublished paper delivered to Stuttgart University in 2014, Woodard speaks of the seasonal generative abilities of plants to blossom, bear fruit and reproduce.[41] This 'growth'

---

39   Laurence, Janet. *The Alchemical Garden of Desire.* McCelland Sculpture Park and Gallery catalogue, Victoria, 2012.

40   Woodard, Ben. *On an Ungrounded Earth: Towards a New Geo-Philosophy.* New York: Punctum Books, 2013, 9.

41   Ibid, 3.

could be read as 'thought,' plant capacities as conscious enactments. He is not suggesting that plant thought or communication is neurological, because plants have no brains within which to have neural activity (though rhizomic systemic intelligence behaves as an equivalent). As Woodard says,

> Various controversy has erupted over whether 'signal integration' – basically a plant's ability to combine various sets of sense data into an action – counts as a form of intelligence. The central cause for why plants do not qualify as having intelligence is due to their lacking neuronal and synaptic structures. In a functional sense plants can communicate to their own bodies, to other plants, and to animal species (wasps to attack caterpillars) through what has been referred to as hormonal sentience.[42]

A second work by Laurence that involves plants and their transformative capacities, in this case young trees, was made by the artist in September 2014. She embarked on a major collaborative project, *The Treelines Track* at Bundanoon, south of Sydney. This natural intervention (planting new native trees along a singular line, with stones laid and inscribed with poetry) was the result of directly working with Landcare Australia's Shane Norrish, and made possible through a Bundanoon Trust grant. The project's genesis began with Laurence walking through the Trust property and wondering about its histories – natural, colonial, pastoral and regenerative. Wombats were a major problem in the area, disrupting new planting with their constant digging, leaving clumsy damage in their wake. Laurence's quote, 'I had to consider the area as a home for the wombats too,' reveals her empathy for animal and plant alike.[43]

To apply Woodard's re-interpretation of Schelling's 'ungrounding' as 'regrounding' to Laurence's work for my own purposes, the disruption of perception of the earth is enacted by the artist in this installation *Treelines Track*. The idea of changing our perceptive realities, via plant life and art life interactions is also in evidence in this work. We consider the bush to be natural. Yet we forget the history of agriculture: of clearing land for pasture; of growing crops, dusting them with chemicals and then leaving paddocks fallow; of colonial litter barely below the soil surface, and so on. In other words, the bush that Laurence re-plants or co-plants is not natural, but artificial. The bushland has a long history of production and interference. In Laurence's work, regrounding is extant because she subverts our utopian experience of the land, she challenges

---

42   Woodard, Ben. "Rootedness and Embodiment," for Unsorcery: Future Nature at Schloss
      Solitude, Stuttgart Germany, July 23, 2014, 3.

43   Laurence, Janet. Interview with the author, Sydney, 4 March 2015.

it, confronts it and redresses our instinctively idealistic associations of the bush. Laurence did this by designing her *Treelines Track* artwork as an address to the history of change in the area.

Michael Marder's Plant-Thought

Laurence's work is directly informed by plant philosopher Michael Marder's concepts of plant-thinking, where plants are afforded the ontological status to move outside their limited role as background/backdrop, *umvelt* or environment. Plants remember and learn, according to the experiments of scientist Monica Gagliano at the University of Western Australia.[44] This confirms what Laurence has instinctively known and represented in her artwork: plants have more functions and capacities than previously thought.

Marder's plant philosophy is an attempt to avoid dichotomies or habitual dyads of man/world, subject/object, plant/landscape. His plant view forms a critique of anthropocentrism. Instead of adopting a solely human lens, he brings attention to the idea of human consciousness being mutually connected to our vegetal roots.[45] With new research proving plants remember and learn to adapt in threatening environments,[46] space is opened up to bestow thinking attributes to plants. This, then, leads to a more equalised view of the hierarchy of all things (a flat ontology) – including plants, animals, desks, bottles or thought itself – where the human is toppled from its apical or dominant position and plants are elevated from their lowly role to a horizontal plane of relevance, metaphorically contradicting the physical verticality of their growth.

The historical activities of using plants for their aphrodisiac, narcotic, sedative and hypnotic qualities should not be abandoned but added to with a new re-interpretation of their relevance, and Laurence pursues this aim. Marder sees plant-thinking as a 'thinking without the head' because human thinking has become increasingly 'de-humanized and rendered plant-like, altered by its encounter with the vegetal world.'[47] These concepts of plant-thinking affect phytology (study of plants) today and effect an aesthetic or artistic interaction

---

44 Gagliano, Monica. "In a Green Frame of Mind: Perspectives of the Behavioural Ecology and Cognitive Nature of Plants." *AoB Plants*, 7 (2014): 5.

45 Marder, Michael. "For a Phytocentrism to Come." *Environmental Philosophy* Vol 11, No 2, 2014: 1.

46 Gagliano, Monica. "In a Green Frame of Mind: Perspectives of the Behavioural Ecology and Cognitive Nature of Plants." *AoB Plants*, No 7, 2014: 4.

47 Marder, Michael. "What is Plant-thinking?" *Klesis: Revue Philosophique*, No 25, 2013: 1.

with plants. 'Thinking like a plant' is what Marder is referring to here, as opposed to referring to plants' ability to comprehend and communicate.

To extend this discussion of plants as intermediary between earth and human, Marder is also concerned with vegetal phenomenology. This is an enquiry into the ontology of sensory plants and a requisite redressing of a perception of their place in relationship to all things. Recent theories regarding the moral and ethical relations between humans and animals have extended to the non-human debates regarding plants.[48] In his article *For a Phytocentrism to Come*, Marder quotes Aristotle: 'Plants seem to live without sharing any locomotion or in perception.'[49] In other words plants only *appear* to live and Marder describes this as a 'vast green blind spot.'[50] It has occurred to Janet Laurence, as well as to Michael Marder, that the vegetal life of plants is an important component in redressing the problems of climate change and environmental damage. The act of plant growth is central to Marder's theories, in his attempts to move beyond the legitimacy of life as being the domain of the animal world alone. By elevating plants as the rightful representatives of sentience, growth and learning, Marder of course risks doing no more than swapping zoocentrism for phytocentrism. Toppling the animal (human) from its apical position in the hierarchy of ontology and replacing it with plants in that super-star role may be no more than a furthering of anthropocentric human hubris.

Is it possible to articulate Laurence's work as an aesthetic of plant experience as a manifestation of the visualisation of plant wonder and of plant information/co-evolution and adaptation during periods of crisis? If the answer is yes, the aesthetic has developed as (and relies upon) an instinctual reaction to critical climate change issues and geo-philosophical urgencies, such as the changes in the strata of the earth that are ripe for ungrounding and regrounding. In an effort to align philosophy with plants, Marder charts various continental philosophers who refer to plants to inform their concepts; for instance, Nietsche's interest in vegetal digestion, Hegel's concession that the act of devouring extends to the nutritive properties of plants, and Freud's repression as an interference with flowering as sexual maturation and human ripeness.[51] When Marder draws a link between 'spirit consciousness' as being only partly

---

48   Marder, Michael. "What is Plant-thinking?" *Klesis: Revue Philosophique*, No 25, 2013.

49   Marder, Michael. "For a Phytocentrism to Come." *Environmental Philosoph,y* Vol 11, No 2, 2014: 5.

50   Ibid.

51   Marder, Michael. "What is Plant-thinking?" *Klesis: Revue Philosophique*, No 25, 2013, 172–175.

exposed to the light, as are plants whose roots are hidden in the dark soil, we see his wish, the urge, the desire to connect plants with 'life.'[52]

These associations between plants, philosophy and life are enduring. The mediation of art in this equation suggests something else. Artworks predicated on the import of plant energy, plant force and plant transformative powers move us into the world of metaphoric ideation. However, a true plant-art ontology must resist the urge to transcend, to idealise and to sublimate. Artists using plants as aggregates in an overall aesthetic system, and plant theorists charting this territory of plant aesthetics, must stand firm and avoid romantic rhetoric. Schelling states that if a certain thing's conditions can't be given in nature, then it is impossible.[53] I contend, as a short summary, that Laurence's art sits both inside and outside nature and must resist transcendence. It is not impossible.

Conclusion

Janet Laurence's work addresses ontological significances in the natural world and her ideas emerge from those questions of where humans sit in relation to other species and other things. The fascination of Michael Marder is his curiosity about a transfigured means of human thought that has become dehumanised and plant-like in its properties, and how that relates to plants. There are random and arbitrary elements of *Treelines Track* in terms of the rogue behaviour of plants and trees. Laurence and Landcare's Shane Norrish can plant new trees but the way they grow, adapt, inhabit, die off or affect other trees is out of the realm of their intentions, as they have no control over the weather, etc. Could tending to vegetal life equate to a tending of human thought? If so this must be one of the highlights of the *plant contract.*

Janet Laurence has been preoccupied with these concepts of finitude and decay, of re-growth and replenishment, for all of her working life. Laurence unearths her plants and risks a topography of hell. Ben Woodard says, 'hell is an all too familiar landscape, often a volcanic region with roaming demons or floating rivers of souls more or less terrestrial.'[54] Woodard agrees with Jan Zalasiewicz when the latter proposes that the earth will bury humans and all

---

52   Ibid 173.

53   Hamilton Grant, Iain. *Philosophies of Nature After Schelling.* London: Continuum, 2006.

54   Woodard, Ben. *On an Ungrounded Earth: Towards a New Geo-Philosophy.* New York: Punctum Books, 2013, 71.

trace of us, as a strange xeno-archaeology.[55] It is as though Laurence is pre-empting this darkly perverse concept of the surface of the earth enfolding us all into its layers as a speculative horror. She intuitively knows that human (and animal and plant) life may be exterminated, or at least that it exists on borrowed intergalactic time.

Edourdo Kohn says, 'To engage with the forest on its own terms, to enter its relational logic, to think with its thoughts, one must become attuned to these.'[56] The connection between art and the biology of plants converges with thought. The power and force of plant life, which extends from medicine chests to Janet Laurence's *Elixir Bar* and her return to a treeline through the Australian scrub, is the entelechy that can alter our consciousness and hopefully our attitude to vegetal life. This is the poroposition of the *plant contract*.

Humans have made plants instrumental by eating them and using them as curative ingredients. Laurence has manipulated plant life, to bring awareness to environmental issues, yet she too has interfered with 'natural' processes. None of us is guiltless. Plants are a source of great sustenance for humans and whilst this book is a turn towards plants in a non-instrumental context, it would be futile to pretend this plant-utility does not exist. It would also be worthless to only perceive plants as benevolent, withholding our intuition that there may be a malevolent side to their secret lives. Does this acknowledgement of the 'whole plant' do no more than represent a slippage into science fiction or does it reveal an understanding of the aggregate properties of plants? The magical possibilities of plants have been discussed here in terms of their capacity to transform human experience but these metamorphic plant qualities may do more than alter the human, they may help reframe conventional views of plants as discrete and inert. This is the question: who gets to cast the vegetal spell? Or should I say: what gets to cast the vegetal spell?

---

55   Zalasiewicz, Jan. *The Earth After Us: What Legacy will Humans Leave in the Rocks.* New York: Oxford University Press, 2008, 67.

56   Kohn, Edourado. *How forests think: Toward an Anthropology Beyond the Human.* Berkeley: University of California Press, 2013, 20.

Conclusion: On Rhizomes and Dead Trees

Ways of Seeing Dead Trees

Years ago, when I was at my cousin's farm at Coolah in Central West NSW, I was sent on a mustering job to bring back four cows from the far border, on horseback. I was fifteen years old and very pleased to have been given the responsibility of this task. It was stinking hot and the grass crunched white under the horses' hooves. The cicadas were so loud, the sound in my ears began to widen until my whole head was throbbing in time with their chant. The saddle's buckle burned against my calf.

After about a half hour, riding up over the ridge of ancient granite stones, lightning struck over the far hill and moved closer. Wind picked up and a few cracks of thunder made me, and my horse, skittish. As I rode along a creek bed and moved up towards a gate at the top of the next hill, a gum tree was struck by lightning and split. At that exact moment that the lightning flashed, I saw an x-ray view of the tree. Inside it, was a human figure, a skeleton.

As the clouds and lightning moved away again, quite quickly, having brought not a single drop of rain, I urged my horse up to the tree. I wanted to see the dead body. But when I got to the top gate, I saw that the tree was completely intact, not split open with a skeleton to be seen inside at all. However, the tree was dead. No foliage, dried out branches, no more than a glorified stick.

The human mind plays ocular tricks on us: we are susceptible to what we want or loathe to see. Our visual perception can be affected by our fears and hopes, and the creative possibilities of our imaginary realms are an interesting unknown. In terms of our perception of nature, this ability to see in a distorted way has affected our ability to see disused vegetal places as beautiful and our ability to understand the truth of other species beyond our human version of them. However the state of play has changed. The dead tree has a significance now that is different.

This book has focussed on plants. Plants that have the right to rights, plants that tease and transform, plants that suggest a better model for multi-gender living, plants that are incorporated into aesthetic motifs lasting over 2000 years and plants that connect us to the soil beneath our feet and the sky above our heads. I have saved up a short discussion of the tree for this final chapter, my conclusion. This is because the tree is the epitome of the plant. The tree resides over my *plant contract*.

A tree is a clear denotation of nature for humans. It is the exultant metaphor for all vegetal life and humanity's downfall. It has, in human terms, gravitas

© KONINKLIJKE BRILL NV, LEIDEN, 2018 | DOI 10.1163/9789004360549_009

and grandeur. So far, I have discussed plant life with respect to life and breath, and thought, by turning to Michael Marder's excrescence and Luce Irigaray's efflorescence. The entire text has been an ode to Michel Serres' natural contract, as a wish to see it persevere. I can't think of a better way to finish than with the image of the dead tree. This is not the end, not death necessarily, but a finitude that never finishes. Post-finitude. The fallen tree decays and turns to dust, turning to compost for the next lease of life. Death is one part in the cycle of life. We see the body of the dead tree and sometimes its stump if it is cut down. But less frequently do we see the rhizomic rooted matter beneath the earth's cool surface.

During the Romantic traditions of the Enlightenment, a dead tree was a framing device used by such painters as Poussin or Constable. A host of Australian Picturesque colonial painters (the aesthetic of the semi-wild) from John Glover to Eugene Von Guerard also used the fallen dead tree across the foreground or mid-ground of the painting to focus the eye as it moves towards the distant background. Often a tree on either side of the picture plane served as a proscenium arch to create a 'mise en scene' in these 19th century paintings. Humans have literally (and conceptually) used nature as a framing device. It's part of our construction of nature that we have seen it as an inert backdrop to human action: we can see this perception in the way we represent it in art and literature.

Watching the development of the dead tree across human (western culture) time is interesting as a means of mapping the wasted and wasting land we live in. The *plant contract* is intended as a story of human relations with the natural world – a strong and endlessly growing image of resilience. From creating a form of governance among humans, to understanding the impact of the human on the world, is to finally understand that a plant contract is a means of opening up deeper discussions about human relations with plants. This becomes its ethical consideration.

If I were to pair a tree artwork with each chapter of this book then the tree that best represents the wasteland concepts of this book is Anselm Kiefer's 2006 *Palm Sunday* where a dead palm tree lies on its side in the gallery.[1] The dead tree, like the winter tree, is an abandoned tree. Abandoned by the heat of summer and the drenching white light that accompanies it. Kiefer's tree has lost all sense of itself, a cruel reminder of industry, death, heartbreak. It reminds us of the cycles of nature, with humans. Trees compete and cooperate, and ultimately fall down with a majestic death. Giuseppe Penone's *Tree of*

---

1   http://www.tate.org.uk/art/artworks/kiefer-palm-sunday-ar00038.

12 Metres 1980–82 is an example of plant-art ethics gone wrong. He pared back fallen trees to their stumpy branches and cut back to their truncated form, to show its previous state of growth. These are distressing sculptures, exhibited at the Tate in London. A wasteland without beauty or a wasteland that becomes beautiful again.[2] Constantin Dimopoulos is an Australian artist who has turned trees blue across Canada and Australia. He was interested in the disjunction between the colour blue and the colours we see in nature.[3] Nature revoked, nature wasted, nature re-performed.

The hybrid tree can stand in for my chapter on hybrid plant human forms, such as the Green Man. The hybrid tree is also a forming tree. US artist Sam Van Aken created the *Tree of Forty Fruit* 2014, where he grafted myriad peach, plum, apricot, nectarine, cherry and almond trees into one. This extreme bounty, like the Green Man motif, has an abundance and allure that buys into our wildest desires. The indulgence of forty different fruits! Yet there is a sense of bizarre and fearful loathing, a monstrous overloading to the Forty Fruit tree that is a magnetic reminder of the motifs of human plant hybrids to date and the human extremism that has led to our ecological deteriorations. What kind of hybridity can we future forecast?

In terms of my robotany chapter, Brandon Ballengee's *Dying Tree* 2013 (an ode to Bruce Nauman's *Amplified Tree* 1970) is surely the best masthead I could hope for.[4] Ballangee's idea was to focus our viewing attention on the process of an ill tree in its last gasps. According to the artist, he placed sensitive microphones into the outer layers of the tree, to amplify the sound of the water evaporating from the wood tissue. Whether the recordings were water evaporating or the sound of other mechanical and vascular processes as hard to ascertain.

Robert Voit, too, created mobile phone masts, disguised as trees, in 2003. The masts simulate nature as a way to remind us of hazardous electrical smog.[5] Finally, along with the disguised media towers and the tortured bonsais is *The Dark Forest,* 2009 which is a project between tropical forests in Brazil with temperate forests in the UK. Sherwood Forest and the Amazon, so disparate but connected by sensors which mobilise the sounds of tracked changes in the forests to one another.[6] These technologized trees are endemic of the third media revolution where we are so consumed by technology that we can no longer distinguish between realities. The 'internet of things' includes plants of

---

2   http://www.tate.org.uk/art/artworks/penone-tree-of-12-metres-t05557.
3   http://kondimopoulos.com/the-blue-trees/.
4   http://brandonballengee.com/dying-tree/.
5   http://robertvoit.com/bilder/serie1_new_trees.
6   http://thedarkforest.tv/index.html.

all kinds (roboticised, augmented, made virtual, inter-activated). This is suggestive of future forms of communication that mimic plant life.

My eco-punks – the artists, writers, activists and plants – have a habit of dismantling and collapsing our intuitive thinking. Irigaray suggests that an absence of plant life marks an absence of thought. Robert Smithson's 1969 tree was an act of eco-punking. He exhibited a dead tree at the Prospect exhibition of Kunsthalle in Dusseldorf of that year. Posthumous reconstructions were made in 1997 and remade in 1999.[7] The massive tree was laid down, roots and all, on the gallery floor with mirrors. Sam Durant's *Upside Down Pastoral Scene* also comprised upended tree roots and mirror and was a reiterated version of Smithson's.[8] Durant's twelve fibreglass stumps, with real roots and branches attached to them, emerge from the gallery floor as a testament to Smithson's conceptual practice. Enrique Oliviera has made a related career from his twisted disappearing trees within gallery spaces, emerging from the floor and escaping through the ceilings, a cosmos of tree and paint, biotic and abiotic existence.[9] His tree works are epic, eco-punk, theatrical implosions of trees and uses plywood to make these bulky protrusions through the gallery floor and shopfronts.[10]

An example of feminist tree art, as a spokesperson for my feminist water lily chapter, must fall to the US based artist Natalie Jeremijenko who created *XClinic Farmacy* which was an installation at the Postmasters Gallery, New York, in 2011, where she hung bags of soil with plants growing in them on the side of a warehouse. These ag-bags were part of what she calls a clinical trial (researching urban agriculture) and were available for purchase from the gallery.[11] This commitment to raising environmental awareness in a social/art context is a thread through all Jeremijenko's work. She also created *Plant Logic* 1999 where she hung plants upside down along a streetscape near the Museum of Contemporary Art, Massachusetts.[12] Women making art alongside plants or utilising plants or communicating with plants, to repeat the communication structures of the vegetal world, is a feminist act. This work with plants may constitute a new way of distributing ideas about feminist ideology.

---

7   https://www.robertsmithson.com/sculpture/dead_tree_300.htm.

8   http://www.afterall.org/journal/issue.10/speaking.others.

9   http://www.henriqueoliveira.com/.

10   http://www.henriqueoliveira.com/portu/depo2_i.asp?flg_Lingua=1&cod_Depoimento
        =36.

11   From database of Curating Cities, a database of eco public art, UNSW. http://eo-publicart
        .org/xclinic-farmacy.

12   http://www.expandedenvironment.org/tree-logic/.

And finally, as a mainstay of ungrounding in my last chapter – as a means of changing our value system relating to perceptions of nature – is my introduction of Joseph Beuys' famous 1982 art installation *7000 Oaks* which was the ultimate ungrounding. Beuys believed art was a means of social utopia and planted 7000 trees through the city of Kassel alongside a basalt stone.[13] The only way to extend an ungrounding any further must surely be a turn to outer space. Shen Shaomin created a land-based bonsai series where he trained bonsai trees into painful and unbearable shapes and positions, to remind humans of what we really do when we engage in gardening. Control, manipulation, augmenting: without thought for the bonsai itself. Shaomin's bonsai's are tortured, using grates, violent clamps and wrenching apparatus, tools to coerce and torture.[14] But it was the artist Azumo Makoto who launched a series of bonsais into space in 2014.[15] This was ungrounding at an extreme level.

These plant-related artists introduced here as a means of concluding and opening up discourse for a future book, have worked with trees (as opposed to other types of plant life). Their praxes tell a story about the disused and the reclaimed, as well as the metaphoric value of the archetypal tree form. They support the theory that there is a movement of environmental art that relates specifically to plants, but in this instance to trees. They also create a landscape of art that slowly changes the conventional aesthetic of art (with a set of previous criteria) towards a genre of plant art that has different modes of critical judgement.

Cross-Species Rhizomes

From primal forms of arborescence, I must at last turn my attention down to the ground, to dig my roots into this area of study. So, from the arbors to the rhizomes. Underground rhizomes or creeping rootstalks can survive for thousands of years, even when the surface trunks and foliage have been devastated by fire or drought, insect attacks or fungus. The rhizome is a source of longevity, fecund with creative possibility. In asexual plant reproduction the rhizome can act as a reproduction system. The tips of the underground or underwater roots can break off as new plants. This is one means of reproduction in water

---

13   http://www.tate.org.uk/art/artworks/beuys-7000-oak-trees-ar00745.

14   http://www.designboom.com/art/shen-shaomin-bonsai-series/.

15   http://www.designboom.com/art/azuma-makoto-exobiotanica-project-bonsai-tree
      -07-21-2014/.

lilies, couch grass and nettles. The attraction of the rhizomic model in plants is their subterfuge, their activity away from human eyes.

Deleuze and Guattari, in *A Thousand Plateaus*, created a system of knowledge that works against arborial concepts which affirm the vertical and hierarchical structures over the roots. Instead, they adopt the allegory of the rhizome as a system of multiplicity, with trans-species connections. I love their opening lines: 'A first type of book is a root book. The tree is already the image of the world, or the root the image of the world-tree.'[16] This book, too, has been an aggregate of fragments, and was intended as an overall schemata of plant-art ideas.

Deleuze and Guattari's evocation of a rhizomic system of multiplicity includes every subterranean part being connected with anything else. This is different, Deleuze and Guattari say, from real tree roots which have a plotted course, a fixed order.[17] This distinction of the rhizomic model from the real rooted model of the plant can be obfuscating in the context of Critical Plant Studies. In fact, plants are more connected, heterogeneous and un-plotted than Deleuze and Guattari give credit for. If studied more closely, the plant root system is constantly shifting and exploring in random searches for nutrients and water. Some roots break free of the crust of the world to support upper branches, and even extend for miles in contingent and speculative searches for life.

The interaction and interactivity between species, biotic and abiotic, might engender our understanding of art and technology, and vice versa. A non-hierarchical model of relevance in the natural world can help humans to re-organise or redress the way we behave towards and within the environment. The most important element of Deleuze and Guattari's rhizomic concept is its non-hierarchical properties. We have seen the devastating results of unsustainable human behaviour such as overuse of resources and mismanagement of the environment – resulting in land erosion and subsidence, chronic carbon exo-flows, disastrous weather and bushfire damage.

Instead of distinguishing a concept of rhizomic knowledge as similar to (but separate from) real plant root systems, Stefano Mancuso writes of the 'swarm of roots.'[18] Rhizomic qualities and their intuitive patterns of behaviour can be linked to properties of flocks of birds. Birds, even in murmuration, seem to fly in all directions without colliding. Each bird relates their flight plan according to the bird nearest them. There seems to be a route the swarm will take

---

16   Gilles Deleuze and Felix Guattari, 'Rhizome' in *One Thousand Plateaus: Capitalism and Schizophrenia*, Minneapolis: University of Minnesota Press, 1987, 5.

17   Ibid.

18   Stefano Mancuso, *Brilliant Green,* New York: Island Press, 2015.

across the sky, often changing direction quickly without mishap. Plants, says Mancuso, concentrate their multiplied and connected behaviour together in a comparative way. This is emergent behaviour, Mancuso says, like the spontaneous clapping amongst an audience of humans.[19] For plants, there is a coordination of activities amongst various numbers of plants or forests, which allows maximum possibility of survival (such as emitting chemicals as warnings). Plants work together in these swarm-like ways, in multiplied and aggregated forms that need further investigation. There is no need for a higher volition or an authority position. While Deleuze and Guattari used rhizomes to illustrate their ideas of how to live economically, politically and socially together, plants are in fact already functioning for this same purpose.

As Mancuso says, 'Like the tips of a root system or ants in a swarm, they amount to nothing by themselves, but together develop incredible capacities ... But in plants, these dynamics actually come into play inside one plant, between its roots. In short, every single plant is a swarm!'[20] It is interesting to consider that Deleuze and Guattari's theory of the rhizome is an interpretation of language, using the formations of plant roots as a metaphor. Meanwhile, plants undertake their own form of language, outside human capacities to translate.

Conclusion

The capacity of art to translate complex issues and render visible ideas and concepts otherwise difficult to discern, makes it a critical accompaniment to the silent rising of the discourse of plant sentience and vegetal thinking. This discourse constitutes a *plant contract*. For this reason, this study has focused on the impact of art on the dissemination of new plant science. The gap between plant life and human life can be filled by the mediatory potential of art, and the artists collected in this book are all working in this lacuna. In facilitating the perception that species are less dissimilar than we knew, their work may be the key to comprehending that 'being' is more than acting as a subject in an objective world. In this way art might even help to reground the geology of the earth, and to shift human perception away from the subjective and towards an aggregated system where neither plants nor aesthetics are cast aside.

---

19   Gilles Deleuze and Felix Guattari, 'Rhizome' in *One Thousand Plateaus: Capitalism and Schizophrenia*, Minneapolis: University of Minnesota Press, 1987, 145.

20   Gilles Deleuze and Felix Guattari, 'Rhizome' in *One Thousand Plateaus: Capitalism and Schizophrenia*, Minneapolis: University of Minnesota Press, 1987, 146.

Art, safely moored at the *plant contract* wharf, is a means to show that there are forceful elements in plant species that we cannot comprehend with static thinking and representation. Plants show an astonishing vitality and resourcefulness in their adaptability and methods of finding water and light. The vitality and adaptability of the plant world challenges and affronts our thinking and our representations.

These associations between plants, philosophy and life, when mediated by art, suggest something beyond the science of plant energy and plant force and drive us to metaphoric ideation. However, a robust plant-art ontology must resist abstraction and the urge to transcend, to idealise and to sublimate and affirm the material, the sensorial as well the symbolic life of plants. The challenge for artists using plants as aggregates in an overall aesthetic system, and plant theorists charting this territory of plant aesthetics, is to remain grounded in the plant lifeworld and avoid romantic rhetoric. The plant and technology artworks discussed in this book sit both inside and outside nature and as a result largely resist the rhetoric of transcendence, which keeps the art experience at a distance, rather than being immersive and immediately or physically sensorial.

The connection between art, science and the biology of plants converges with robotanical and ethical thought and combine as a construction of a *plant contract*. Plants understood as rhizomic intelligence systems have an independence (or are pre-programmed computationally as robotic art), yet they retain an entelechy (vital principle guiding development). An awareness of plants as autonomous and agented things fuels a consciousness of anthrodecentric human/nature relationships. Into the future we can see that studies based on such trans-species frameworks will need to account for other emergent forms of non-human life. Just as the vegetal soul seeks nourishment from its environment and works to reproduce itself, so too does artificially intelligent life. Within such a framework, the human brain, the vegetal form and AI will exist together equally. That story is yet to be written. But, here, today? Consider this book my signing of a *plant contract*.

Bibliography

Ackroyd, Heather and Dan, Harvey. "Chlorophyll Apparitions." *Signs of Life: Bio Art and Beyond*. Ed. Eduardo Kac. New York: MIT Press, 2006.

Albrecht, Glenn, et al. "Solastalgia: The Distress Caused by Environmental Change." *Australas Psychiatry* 15 Suppl. 1 (2007): S95–S98.

Aloi, Giovanni (ed). "Beyond Morphology." *Antennae* 18, Autumn (2011).

Bannon, Bryan E. "Re-envisioning Nature: The Role of Aesthetics in Environmental Ethics." *Environmental Ethics* 33, Winter (2011).

Barad, Karen. "Nature's Queer Performativity." *Qui Parle* 19, 2 (2011): 121–158.

Barrikin, Amelia. *Parallel Presents: Pierre Hugye*, New York: MIT Press, 2015.

Bauman, Zygmunt. *Postmodern Ethics*, Oxford: Blackwell Publishing, 1993.

Blamey, Peter. "Unexplored Functions/Everyday Objects: Interview with Tully Arnot." *Das Superpaper*, Issue 33, November 2014.

Bourriaud, Nicolas. *The Radicant* New York: Lukas and Sternberg, 2009.

Brits, Baylee. "Brain Trees: Neuroscientific Metaphor and Botanical Thought." *The Covert Plant* Santa Barbara: Punctum Books, 2017.

Burke, Edmund. *A Philosophical Inquiry into the Origin of our ideas of the Sublime and the Beautiful.* London: Oxford Uni Press, 2015.

Calvo, Paco. "The Philosophy of Plant Neurobiology: A Manifesto." *Synthese*, 5, 193 (2016): 1323–1343.

Carlson, Allen. "Is Environmental Art an Aesthetic Affront to Nature?" *Canadian Journal of Philosophy* 16 (1986).

Casteneda, C. "Robotic Visions." *Social Studies of Science* 44, 3 (2013).

Cavender, Anthony and Albin, Manuel. "The use of magical plants by curanderos in the Ecuador highlands." *Journal of Ethnobiology and Ethnomedicine* 5, 3, 2009.

Centerwall, Brandon. "The Name of the Green Man." *Folklore* 108 (1997).

Chamovitz, Daniel. *What a Plant Knows*, NY: Scientific American/Farrar, Straus and Giroux 2012.

Colebrook, Claire. *Sex After Life*, Open Humanities Press, 2013.

Corner, James, Scofido, Diller and Renfro. *The High Line: Foreseen, Unforeseen*, London: Phaidon, 2015.

Crisinel, Anne-Sylvie. Spence, Charles. Cosser, Stefan. Petrie, James. King, Scott. Jones, Russ. "A bittersweet symphony: Systematically modulating the taste of food by changing the sonic properties of the soundtrack playing in the background." *Food Quality and Preference* 24, 1 (2012): 201–204.

Cullinan, Cormac. *Wild Law: A Manifesto for Earth Justice*. Devon: Green Books, 2011.

Darlington, Susan. *The Ordination of a Tree: The Buddhist Ecology Movement in Thailand,* Albany: SUNY Press, 2012.

Darwin, Charles. *The Movement of Plants,* London: John Murray, 1880.

Deleuze, Gilles and Guattari, Felix. 'Rhizome' in *One Thousand Plateaus: Capitalism and Schizophrenia*, Minneapolis: University of Minnesota Press, 1987.

De Wachter, Ellen Mara. "Art and Life." *Frieze*, 24 April 2016.

Dinshaw, Carolyn. 'Black Skin, Green Masks: Medieval Foliate Heads, Racial Trauma, and Queer World-making'. *The Middle Ages in the Modern World*. Ed. Bettina, Bild-hauer and Chris, Jones. Oxford: Oxford University Press, 2017.

Doyle, Richard. *Darwin's Pharmacy: Sex, Plants and the Evolution of the Noosphere,* Washington: University of Washington Press, 2011.

Dudareva, N. Klempien, A. Muhlemann, J.K. Kaplan, I. "Biosynthesis, function and metabolic engineering of plant volatile organic compounds." *New Phytologist* 198, (2013): 16–32.

Firn, Richard. "Plant Intelligence: An Alternative Point of View." *Annals of Botany* 93 (2004): 345–51.

Funnell, Antony. "The Underestimated Power of Plants." *Future Tense*, ABC Radio National 8 March 2016, http://www.abc.net.au/radionational/programs/futuretense/the-underestimated-power-of-plants/7227008.

Gagliano, Monica. "Experience teaches plants to Learn Faster and Forget Slower in Environments Where it Matters." *Oceologia* 175 (2014).

Gagliano, Monica. "In a green frame of mind: perspectives on the behavioural ecology and cognitive nature of plants." *AOB* 7, (2015).

Gandy, Matthew. "Marginalia: Aesthetics, Ecology, and Urban Wastelands." *Annals of the Association of American Geographers*, 6, 103 (2013).

Gibson, Prudence (ed). "Interview with Michael Marder" in *The Covert Plant,* Santa Barbara: Punctum Books, 2017.

Gibson Prudence. *Janet Laurence: The Pharmacy of Plants.* Sydney: New South Press, 2015.

Gibson, Prudence. "Climate Cry" and "The Underground Garden." *Climate Century*, Vitalstatistix, Adelaide, 2015.

Gibson, Prudence. 'Pavlov's Plants', *The Conversation*, 6 December 2016. https://theconversation.com/pavlovs-plants-new-study-shows-plants-can-learn-from-experience-69794. Accessed 6 January 2017.

Goethe, Johann Wolfgang Von. *Italian Journey,* London: Penguin classics, 1962.

Goethe, Johann Wolfgang Von. *The Metamorphosis of Plants,* London: MIT Press, 2009.

Grosz, Elizabeth. *Chaos, Territory, Art.* New York: Columbia University Press, 2008.

Guattari, Felix and Deleuze, Gilles. *A Thousand Plateaus,* Minneapolis: Minnesota Press, 1987.

Hamilton Grant, Iain. *Philosophies of Nature After Schelling.* London: Continuum, 2006.

Hall, Matthew. *Plants as Persons: A Philosophical Botany*, Albany: SUNY Press, 2011.

Hall, Matthew. "Plant Autonomy and Human-Plant Ethics." *Environmental Ethics* 2009.

Haraway, Donna. *When Species Meet*, Minneapolis: Minnesotta Press, 2007.

Haraway, Donna. "Anthropocene, Capitalocene, Plantationocene, Chthulucene: Making Kin." *Environmental Humanities*, 6 (2015): 159–165.

Haraway, Donna. "Sowing Worlds: A Seed Bag for Terraforming with earth Others" in Grebowicz, Margaret and Merrick, Helen. *Beyond the Cyborg: Adventures with Haraway*, New York: Columbia University Press, 2013.

Hayman, Richard. *The Green Man*. Sussex: Shire Publications, 2015.

Hicks, Clive. *The Green Man: A Field Guide*. Fakenham: Compass Books, 2000.

Hirthe, G. Porembski, S. "Pollination of *Nymphaea lotus* (Nymphaeaceae) by rhinoceros beetles and bees in the northeastern Ivory Coast." *Plant Biology* 5, (2003): 670–676.

Houle, Karen. "Animal Vegetable, Mineral: Ethics as Extension or Becoming: The Case of Becoming Plant." *Journal for Critical Animal Studies*, ix, 1–2, 2011.

Hughes, Robert. *The Art of Australia*. Melbourne: Penguin, 1966.

Irigaray, Luce and Michael, Marder. *Through Vegetal Being*. New York: Columbia University Press, 2016.

Irigaray, Luce. *I Love to You*, New York: Routledge, 1996.

Irigaray, Luce. *This Sex Which is Not One*. Ithaca: Cornell University Press, 1985.

Irigaray, Luce. "Animal Compassion" in M. Collarco and P. Atterton (eds) *Animal Philosophy: Essential Readings in Continental Thought*, London: Continuum, 2004.

James, Jamie. "W.B. Yeats: Magus," *Lapham's Quarterly*, Summer (2012), accessed August 12, 2013, http://www.laphamsquarterly.org/essays/wb-yeats-magus.

Kac, Edourdao. *Signs of Life: Bio Art and Beyond*. Ed. Eduardo Kac. New York: MIT Press, 2006.

Kac, Eduardo. "Foundation and Development of Robotic Art." *Art Journal*, 56, 3, Fall (1997).

Koechlin, Florianne. "The Dignity of Plants." *Plant Signaling and Behaviour* 4, 1 (2009).

Kohn, Edourado. *How forests think: Toward an Anthropology Beyond the Human*. Berkeley: University of California Press, 2013.

Kramer, H. *Malleus Maleficarum*. Speyer, Germany 1487.

Kristeva, Julia. *Powers of Horror: An Essay on Abjection*. New York: Columbia University Press, 1982.

Laist, Randy (ed). *Plants and Literature: Essays in Critical Plant Studies*. New York: Rodopi, 2013.

Lingis, Alphonso. *The First Person Singular*, Northwestern University Press, 2007.

Mabey, Richard. *The Cabaret of Plants*. London: Profile Books, 2015.

Mancuso, Stefano. *Brilliant Green*, Bologna: Island Press, 2013.

Marder, Michael. *Plant Thinking: A Philosophy of Vegetal Life*. New York: Columbia, 2013.

Marder, Michael. *The Philosopher's Plant: An Intellectual Herbarium*. New York: Columbia University Press, 2014.

Marder, Michael. "Resist Like a plant! On the Vegetal Life of Political Movements." *Peace Studies Journal* 5, 1 (2012).

Marder, Michael and Anais Tondeur, Anais. *The Chernobyl Herbarium*, Open Humanities Press, 2016.

Marder, Michael. "What is Plant-thinking?" *Klesis: Revue Philosophique*, 25 (2013): 124–143.

Marder, Michael. "Should Plants Have Rights?" *TPM* 3rd Quarter 2013.

Marder, Michael. "For a Phytocentrism to Come." *Environmental Philosophy* 11, 2, 2014.

Marder, Michael. "Do Plants have their own form of consciousness?" *Al Jazeera,* 25 June 2012, http://www.aljazeera.com/indepth/opinion/2012/06/2012619133418135390.html.

Marder, Michael. "Ethics and morality." *LA Review of Books* web site, http://philosoplant.lareviewofbooks.org/?p=177 Accessed 6 January 2017.

McKenna D.J., Callaway J.C., Grob C.S. (1998). "The scientific investigation of ayahuasca: A review of past and current research". *The Heffter Review of Psychedelic Research* 1: 65–77.

Miller, Elaine. *The Vegetative Soul: From Philosophy of Nature to Subjectivity in the Feminine*, Albany: SUNY Press, 2002.

Moore, A. Bissete, S. and Totleben, J. *Saga of the Swamp Thing*, Vertigo, 1987.

Morton, Timothy. 'This Biosphere Which is Not One: Towards Weird Essentialism,' *The Journal of the British Society for Phenomenology,* 46, 2 (2015).

Morton, Timothy. *Dark Ecology: For a Logic of Future Co-existence.* New York: Columbia University Press, 2016.

Muecke Stephen, 'On Magical Language; Multimodality and the Power to Change Things,' in Margit Bock and Norbert Pachler, eds., *Multimodlity and Social Semiosis: Communication, Meaning-Making, and Learning in the Work of Gunther Kress* New York: Routledge, 2013.

Murphie, Andrew. "Convolving Signal: Thinking the Performance of Computational Processes." *Performance Paradigm*, 9, (2013).

Myers, Natasha. "From Edenic Apocalypse to Gardens Against Eden." *Infrastructure, Environment, and Life in the Anthropocene*, Duke University Press, forthcoming 2017.

Myers, Natasha and Hustak, Carla. "Involuntary Momentum: Affective Ecologies ad the Sciences of Plant/Insect Encounters." *A Journal of Feminist Cultural Studies*, 23, 3 (2012).

Nassar, Dalia. "Metaphoric Plants: Goethe's *Metamorphosis of Plants* and the Metaphors of Reason." *The Covert Plant.* Santa Barbara: Punctum Books, 2017.

Nealon, Jeffrey. *Plant Theory: Biopower and Vegetable Life*. Stanford: Stanford University Press, 2016.

Negus, Tina. "A Photographic Study of Green Man and Green Beasts in Britain." *Folklore* 114.2 (2003): 247–61.

Newton, Isaac. *Two Incomplete Treatises*. Bettenham, London, 1745.

Opperman, Serpil. "Theorizing Ecocriticism: Toward a Postmodern Ecocritical Practice." *Interdisciplinary Studies in Literature and Environment* 13, 2, Summer (2006).

Parikka, Jussi. *A Geology of Media*. Minneapolis: University of Minnesota Press, 2015.

Parikka, Jussi. *The Anthrobscene*. Minneapolis: University of Minnesota Press, 2014.

Pasco, Bruce. *Dark Emu: Black Seeds*, Broome: Magabala Books, 2014.

Parsons, Glenn. *Aesthetics and Nature*, New York: Continuum, 2008.

Pendell, Dale. *Pharmako Dynamis: Stimulating Plants, Potions and Herbcraft*, North Atlantic Books, Berkeley 2010.

Petric, Spela. 'Confronting Vegetal Otherness'. *Skotopoiesis*, 1. Accessed 6 December 2016 http://www.spelapetric.org/portfolio/skotopoiesis/.

Plumwood, Val. *Feminism and the Mastery of Nature*, London: Routledge, 1993.

Raglan, Lady Julia. "The Green Man in Church Architecture." *Folklore* 50.1 (1939).

Ranciere, Jacques. *Aesthesis: Scenes from the Aesthetic Regime of Art*, London: Verso, 2013.

Ring, Heather. "The Dignity of Plants" *Archinect* 23 March (2009) http://archinect.com/features/article/86646/the-dignity-of-plants Accessed 20 July 2015.

Rousseau, Jean Jacques. *The Social Contract*, 1762. https://www.ucc.ie/archive/hdsp/Rousseau_contrat-social.pdf Accessed 6 January 2017.

Sekowska, Elzbieta. "Meet the world's Largest Living Organism". *IFL Science*. http://www.iflscience.com/plants-and-animals/meet-worlds-largest-living-organism/.

Serres, Michel. *The Natural Contract*. Ann Arbor: University of Michigan Press, 1995.

Serres, Michel. *Biogea*, Minneapolis: Univocal, 2012.

Serres, Michel. "Faux et Signeux de Brume: Virginia Wolf's Lighthouse." *SubStance* 37, 2, 116 (2008).

Shtier, Ann. "Botany in the Breakfast Room: Women and Early Nineteenth Century British Plant Study" in Pnina, Abir-Am and Dorinda, Outram (eds) *Intimate Lives: Women in Science1789–1979*, London: Rutgers University Press, 1987.

Simard, Suzanne. "Leaf Litter, Expert Q and A." *Biohabitats*, 15, 4 (2016). http://www.biohabitats.com/newsletters/fungi/expert-qa-suzanne-simard/ Accessed 6 January 2017.

Smite, Rasa; Smits, Raitis; Ratniks, Martins. *Talk To Me: Exploring Human-Plant Communication*, RIXC, Riga, 2014.

Smith, Bernard. *Place, Taste and Tradition*. Sydney: Ure Smith, 1945.

Sommerer, Christa and Mignonneau, Laurent. *Interactive Plant Growing: A-Volve*, Tokyo: InterCommunication Center, 1994. http://www.interface.ufg.ac.at/christa-laurent/WORKS/FRAMES/FrameSet.html.

Stewart, Amy. *Wicked Plants: The Weed that Killed Lincoln's mother and Other Botanical Atrocities*, Algonquin Books, Chapel Hill, 2009.

Stone, Christopher. "Should Trees Have Standing: Towards legal rights for Natural objects." *South Californian Law Review*, 45, (1972): 450–501.

Strauss, Levi. "Interview with Michael Taussig; The Magic of the State," *Cabinet magazine*, 18, Summer (2005).

Struik, P., Yin, X., Meinke, H. "Perspective Plant Neurobiology and green Plant Intelligence: Science, Metaphors and Nonsense." *Journal of the Science of Food and Agriculture*, 88 (2008): 363–370.

Taylor, Paul W. "The Ethics of Respect for Nature." *Environmental Ethics*, 3, 1981.

Taussig, Michael. *The Nervous System* New York: Routledge, 1992.

Taussig, Michael. *The Magic of the State*, Routledge, New York, 1997.

Tomkins, Peter. *The Secret Life of Plants,* Harper Collins, New York, 1989.

Trewavas, Anthony. "Aspects of Plant Intelligence." *Annals of Botany*, Vol 92, No 1 (2003): 1–20.

Trewavas, Anthony. "Response to Alip et al: Plant Neurobiology – all metaphors have value." *Trends in Plant Science*, 12, 6 (2007).

Tupper, Kenneth (January 2009). "Ayahuasca Healing Beyond the Amazon: The Globalization of a Traditional Indigenous Entheogenic Practice". *Global Networks: A Journal of Transnational Affairs*. 9 (1): 117–136.

Uexküll, Jakob Von. "A Stroll Through the Worlds of Animals and Men: A Picture Book of Invisible Worlds." *Instinctive Behavior: The Development of a Modern Concept*, ed. and trans. Claire H. Schiller. New York: International Universities Press, 1957.

Underhill, Nancy. *Sidney Nolan*, Sydney: New South Books, 2015.

Van Der Veen, Marijke. "The Materiality of Plants: plant-people entanglements." *World Archaeology*, 46, 5 (2014): 799–812.

Warren, Karen (ed). *Ecological Feminist Philosophies* Bloomington: Indiana University Press, 1996.

Wiersema, J.H. "Reproductive biology of *Nymphaea* (Nymphaeaceae)." *Annals of the Missouri Botanical Garden* 75, (1988): 795–804.

Wohlleben, Peter. *The Hidden Life of Trees,* Back Inc, Sydney, 2015.

Woodard, Ben. *On an Ungrounded Earth: Towards a New Geo-Philosophy.* New York: Punctum Books, 2013.

Woodard, Ben. "Rootedness and Embodiment." for Unsorcery: Future Nature at Schloss Solitude, Stuttgart Germany, July 23, 2014.

Wulf, Andrea. *The Invention of Nature*. London: John Murray, 2015.

Zalasiewicz, Jan. *The Earth After Us: What Legacy Will Humans Leave in the Rocks.* New York: Oxford University Press, 2008.

Zylinska, Joanna. 'Bioethics' in *Telemorphosis:Theory in the Era of Climate Change* Vol 1, Open Humanities Press, 2012.

Index

Printed in the United States
By Bookmasters